Lectures
on fluid mechanics

Marvin SHINBROT

Dover Publications, Inc.
Mineola, New York

Bibliographical Note

This Dover edition, first published in 2012, is an unabridged republication of the work originally published in 1973 by Gordon and Breach, Science Publishers, Inc., New York.

International Standard Book Number
ISBN-13: 978-0-486-48817-2
ISBN-10: 0-486-48817-9

Manufactured in the United States by Courier Corporation
48817901
www.doverpublications.com

This book is dedicated to the heroic Vietnamese people

Preface

THIS BOOK developed out of a course of lectures I first gave at Northwestern University in the summer of 1967. My purpose, then as now, was to introduce the mathematically sophisticated listener to some of the problems of the mechanics of an incompressible fluid. My feeling is that many mathematicians and graduate students are intrigued by what they hear about such problems but don't study them because they don't know enough physics and fear to make fools of themselves. What they don't know, I think, is how abysmally little we actually know about fluids, and how it would be hard to act more the fool than many have already done. I believe the difficulty arises first, over an inflated nomenclature that burdens the subject and, second, over a lack of understanding about how ignorant we can be in our technological society and of how close to the surface so many problems lie.

This belief, coupled with the knowledge of my intended audience, determined the format of the book. It begins, for example, with a derivation of the equations of fluid motion from statistical mechanics. This derivation, while not the usual one, has the virtue of requiring the minimal knowledge of physics (only that force equals mass times acceleration), although it creates some computational difficulties. I was willing to accept these since it is my experience that many students who have little background in physics often have great tolerance for mathematical manipulation. The derivation from statistical mechanics also allowed me to demonstrate at a very early stage how little we know, and to introduce in the very first chapter interesting and important problems in the field.

The rest of Part I is devoted to what is now thought of as the classical part of the theory, most of it developed in the eighteenth and nineteenth centuries. It was necessary to introduce this material since, having put no physics in at the start, no physical insight is available at the end unless one is able to simplify the situation sufficiently that one can draw pictures of flows and so grasp that what is being talked about has something to do with what are ordinarily thought of as fluids. The material in Part I also allowed me to cover the nomenclature that many students appear to find confusing, and again to discuss briefly a number of profound unsolved problems.

In Part II, I cover a small part of the modern mathematical theory of a viscous, incompressible fluid, much of it based on my own work. This is the material that I think mathematicians want most to understand. Here again,

vii

my attitude is that our ignorance is abysmal, and that in spite of the profound methods we use to attack its problems, we know very little about fluids, we can tell the physicist almost nothing of what he wants to know, and interesting problems abound.

Although the considerations discussed above determined the general format of the book, they of course did not determine the specific material covered. That was determined by my own interests, and no one should think that the book covers most of what is of interest about fluids. Compressible fluids, for example, are hardly discussed at all. The reason for this, as I said, is my own interest, and this has been determined by my desire to study fluids in so simple a context that it appears at least possible to understand some portion of the theory down to the ground. Possible it may appear, but anyone who knows the subject is aware that I and everyone else who has tried have so far failed in this endeavor.

The amount of material included seems to be slightly more than what one can cover in a year's course of lectures. That it is a little more is due to the fact that at one time or another when I have given these lectures I have not felt like covering some bit of material and have replaced it by another. Another lecturer may wish to cover still other material. Possibilities are often given in the references.

It is assumed that the reader is familiar with functional analysis, some complex variables, and little else. In particular, it is not assumed that the reader knows anything about partial differential equations, and anything that is needed from that field is proved in the book.

I should like to thank the National Science Foundation for its support while this book was being written.

M. SHINBROT

Vancouver

Contents

PART I

SETTING THE SCENE

Introduction

THE SCIENCE of mechanics begins, and properly ends, as a part of physics. But I believe there is a stage in the development of each branch of mechanics where it is (or should be) part of mathematics. At least one branch—particle mechanics—has passed through its period as mathematics and has become physics again. But fluid mechanics and, more generally, all continuum mechanics have really just begun their development as mathematical sciences. There is much to be done before continuum mechanics can be considered as having reached such a state of symmetry and elegance that the mathematician can turn it back to the physicist for its final development.[1]

Classical particle mechanics began its transition from physics to mathematics with Newton's treatment of gravitation. Before Newton's time, the basic (and basically physical) ideas such as force, mass, uniform velocity and acceleration had to be explored. In addition, much fundamental physical data had to be collected: the distance of the moon from the earth and the curves traced out by the planets are examples.

Although there is some doubt[2] as to how far Newton carried the mathematical development of particle dynamics, there can be no question that the work he did there was mathematical in nature. In the theory of gravitation, Newton had to begin with Kepler's empirical conclusions about the orbits of the planets and then derive these same conclusions from a set of hypotheses about the world. This process, as we now understand it, involves three steps. In order to see exactly where it becomes mathematics and not physics, it is well to enumerate these steps clearly.

(i)   First, the idea is needed that force and acceleration (whatever *they* are!) are proportional. This fact is, of course, the famous "second law" of Newton:

$$F = ma. \tag{0.1}$$

---

[1] It is unfortunately true, however, that for the most part today's mathematicians have not taken their responsibility to mechanics as seriously as did their counterparts in the eighteenth and nineteenth centuries. A consequence has been that the development of continuum mechanics has remained mostly in the hands of physicists and engineers. But that is another story.

[2] See C. Truesdell, A program toward rediscovering the rational mechanics of the age of reason. Archive for history of the exact sciences, 1 (1960) 3–36.

Although this idea will eventually develop into the mathematical equations of motion of the gravitating system, it is important to notice that the validity of (0.1) is a *physical* fact. There is no intrinsic mathematical reason whatever why this law should not read differently. Indeed, it is implicit in some of Aristotle's essays that force is proportional—not to acceleration—but to velocity. And even now, the books are not closed; it may yet happen that in order to explain certain phenomena the velocity, or perhaps its second derivative, multiplied by a coefficient that is very small in ordinary circumstances, may have to be introduced into (0.1). Be that as it may, Newton's law (0.1) has not yet had to suffer real modification, and of course we shall assume it here, remarking again only that it is a *physical* law.

(ii)   The second step entails the specification of the force F in (0.1) as a function of whatever it may depend on. Newton's law of gravitation asserts that for two mutually gravitating particles F is given by the well known inverse square law. Again, this is a physical fact. Although one can invent heuristic mathematical reasons why one might expect *a priori* that Newton's law of gravitation is valid, it is easy to imagine a universe in which it is not. Indeed, if the gravitating particles are within nuclear distances of each other, it appears that the inverse-square law is violated and another description of the force field may be required.

(iii)   It is in the third step that the theory of gravitating particles became mathematics. First, of course, Newton had to explicate the concept of acceleration. This meant the invention of the derivative, and with it the calculus, which remains to this day a part of mathematics. But even that was not enough. In order to explain Kepler's laws, Newton effectively had to solve the differential equation (0.1) to show, for example, that two gravitating particles move in ellipses about their center of mass. And this also is part of mathematics.

There is more mathematics yet in this two-body central force problem. What comes out of Newton's work is that Kepler's laws follow from his. There is also a converse. Given the second law (0.1), Newton's law of gravitation follows from Kepler's laws.[1] Of course, this is a theorem in mathematics, although it has profound physical and philosophical consequences.

In the hands of the eighteenth and nineteenth century mathematicians Euler, Daniel Bernoulli, Lagrange, Cauchy, Laplace, Riemann, and literally hundreds of others, Newton's theory of gravitation developed into what is now called classical particle mechanics. Although it was not always thus, this subject is no longer taught as part of the mathematics curriculum. It has been

---

[1] For a proof, see Max Born, Natural philosophy of cause and chance, appendix 2. Oxford: Clarendon Press, 1949.

returned to the physicists for integration into, and further development as, physics, and to the engineers for use in technology.

Thus, classical particle mechanics can now be thought of as complete.[1] It is a symmetrical, beautiful body of knowledge that, however it may be added to, can continue to be thought of as forming a coherent whole.[2] And at least one problem in the field has been entirely and satisfactorily solved. None of these things is true of continuum mechanics.

This is not to say that no problems in fluid mechanics have ever been solved. It is to say that the field is characterized by its fragmentation, by the fact that various problems in it are often solved by *ad hoc* methods, and that no aspect of it is understood down to the ground. All this makes it difficult for a student of mathematics to enter the field. At the beginning, such a student is faced with what seems a never-ending list of equations describing some portion of fluid mechanics: the Euler equations, the Stokes equations, the Oseen equations, the Navier–Stokes equations, the boundary layer equations, to mention some of them. Where is he to begin? Which of these is fundamental, and which created to aid in the explication of some specific physical problem? In my experience, most mathematicians, at least, only learn the answers to these questions *after* they have worked on a specific problem in the field presented to them on an *ad hoc* basis. One of my purposes in writing this book is to introduce the mathematics student to some of the terms and sets of equations in the field, in a way that does not leave him feeling alone in a vast sea of terminology and methods that he cannot understand without beginning again as an undergraduate in a different major.

As a technique for achieving this end, I shall constantly point out unsolved problems as they are encountered,[3] and I shall constantly criticize the foundations on which the various theories lie. The three steps outlined above serve as a good beginning for this program. In terms of these three steps, the state of continuum mechanics is this. The first step remains. The equations of motion corresponding to (0.1) are well-established, and there is no reason to believe they require any change for ordinary states of matter.

---

[1] Of course, this does not mean that no one any longer works in particle mechanics, or that there are no interesting problems remaining in the field. People do, and there are.

[2] To convince oneself of this, he has only to read any of the elegant texts on the subject. Two good ones, out of many, are: E.T.Whittaker, Analytical dynamics, New York: Dover 1944; and Herbert Goldstein, Classical mechanics, New York: Addison–Wesley, 1950.

[3] As a consequence, this work can be used as the beginnings of a sourcebook for problems for Ph.D. theses. But the production of such a work was not my main intent.

In the second step, a great deal remains to be done. As we shall see later on, the forces in a fluid are by no means completely understood. However, as in the case of particle mechanics, this is a problem for the physicist. What we can say now is that certain assumptions can be made and that we are assured by the experimentalists that they are consistent with certain models of a fluid. Although these assumptions may not remain the last word on the subject, studies involving them can be made to derive conclusions on the models of matter in question.

For fluid mechanics, the third step is in a disreputable state. For example, the most elementary thing that a mathematician can ever say about a physical problem is that it has only one solution, that the pendulum we set moving today will not next Tuesday afternoon at 5:00 o'clock be able to decide whether it will go forward or backward. Such an assertion can *not* be made for any but the simplest models of a fluid. Thus, in contrast to the situation in particle mechanics, the condition of the mathematical aspects of fluid mechanics is abjectly primitive. It is in the hope of ultimately rectifying this situation that I have addressed these sermons to a basically mathematical audience.

CHAPTER 1

The equations of motion

1 Notation

WE SHALL save time later if some notational questions are cleared up now. Points in R^3 are usually denoted by the letter q, with or without subscripts. We always assume there is a fixed rectangular coordinate system with respect to which the fluid motion is described. The coordinates of a point q with respect to this coordinate system are denoted by x, y, and z; the coordinates of q_1 are (x_1, y_1, z_1), and so on. If $q(t)$ is a moving point in R^3 (t is the time), the velocity of the point is denoted by $s(t)$. The components of s in the fixed coordinate system are denoted by u, v, and w:

$$s = (u, v, w).$$

The Euclidean length of an element $q \in R^3$ is denoted by the usual absolute value symbol, so that

$$|q|^2 = x^2 + y^2 + z^2.$$

The volume element in R^3 is denoted by dq. In accord with this notation, instead of the cumbersome

$$\iiint f(x, y, z) \, dx \, dy \, dz,$$

we write simply

$$\int f(q) \, dq.$$

A *domain* is an open, connected set. If V is a domain, ∂V is used to denote its boundary. The surface integral over the boundary of V is denoted by

$$\int_{\partial V} f(q) \, dA,$$

dA being the area element on ∂V.

Literal subscripts *always* denote partial derivatives. Thus, if f is a function of q, $f_x = \partial f / \partial x$. The gradient is denoted, as usual, by ∇. Thus,

$$\nabla f = (f_x, f_y, f_z)$$

in the given coordinate system. Also as usual, the scalar product of two vectors in R^3 is denoted by a dot; for example,

$$\nabla f \cdot \nabla g = f_x g_x + f_y g_y + f_z g_z.$$

7

All this is for R^3. On the rare occasions (mostly in the next few sections) when we have to work in R^n, the notational conventions just sketched are changed a little. Points in R^n are denoted by Greek letters, ξ, η, ζ. The components of ξ (again in a fixed coordinate system) are denoted by subscripts:

$$\xi = (\xi_1, \ldots, \xi_n).$$

The rest of the notation is the same, except that we write

$$\nabla_\xi = \left(\frac{\partial}{\partial \xi_1}, \ldots, \frac{\partial}{\partial \xi_n} \right).$$

2 The transport theorem

Consider a famlly of transformations in R^n defined by equations

$$\xi = \xi(t, \xi^0), \tag{2.1}$$

depending smoothly on a parameter t, which we think of as time ‚We assume the transformation is just the identity when $t = 0$, so that

$$\xi(0, \xi^0) = \xi^0. \tag{2.2}$$

Let $V(0)$ be a domain that is transformed into a domain $V(t)$ by (2.1). In what follows, we have frequent need for a formula for the quantity

$$\frac{d}{dt} \int_{V(t)} F(t, \xi) \, d\xi. \tag{2.3}$$

This formula, which we now derive, is sometimes called *the transport theorem*.

The transformation (2.1) has a Jacobian attached to it, defined by

$$\frac{\partial \xi}{\partial \xi^0} \equiv \begin{vmatrix} \dfrac{\partial \xi_1}{\partial \xi_1^0} & \cdots & \dfrac{\partial \xi_n}{\partial \xi_1^0} \\ \cdots & \cdots & \cdots \\ \dfrac{\partial \xi_1}{\partial \xi_n^0} & \cdots & \dfrac{\partial \xi_n}{\partial \xi_n^0} \end{vmatrix}. \tag{2.4}$$

Because of (2.2), the Jacobian at $t = 0$ is equal to unity. Since (2.1) is smooth, $\partial \xi / \partial \xi^0$ is certainly positive if t is small enough. In the rest of this section, we argue only on those values of t small enough that

$$\frac{\partial \xi}{\partial \xi^0} > 0. \tag{2.5}$$

To derive the transport theorem, we need a formula for the derivative of $\partial \xi / \partial \xi^0$. Let Ξ_{ij} be the cofactor of $\partial \xi_i / \partial \xi_j^0$ in (2.4). Then, by a well known property of determinants,

$$\sum_k \frac{\partial \xi_i}{\partial \xi_k^0} \Xi_{kj} = \delta_{ij} \frac{\partial \xi}{\partial \xi^0}, \tag{2.6}$$

where δ_{ij} is the Kronecker delta:

$$\delta_{ij} = \begin{cases} 0, & i \neq j, \\ 1, & i = j. \end{cases}$$

The derivative of a determinant is the sum of the determinants obtained by differentiating the columns one at a time. Differentiate (2.4), and evaluate each determinant that appears by expanding along the column in which the extra derivative appears. The result is

$$\frac{d}{dt} \left(\frac{\partial \xi}{\partial \xi^0} \right) = \sum_{j,k} \frac{\partial^2 \xi_j}{\partial t \, \partial \xi_0^k} \Xi_{kj}. \tag{2.7}$$

For fixed ξ^0, (2.1) describes a curve in R^n. Denote its tangent by η, so that

$$\eta = \xi_t \, (t, \, \xi^0). \tag{2.8}$$

Because of (2.5), the map from $V(0)$ to $V(t)$ is one-to-one. Therefore, ξ^0 can be thought of as a function of t and ξ. According to (2.7), we have

$$\frac{\partial}{\partial t} \left(\frac{\partial \xi}{\partial \xi^0} \right) = \sum_{j,k} \frac{\partial \eta_j}{\partial \xi_k^0} \Xi_{kj}$$

$$= \sum_{i,j,k} \frac{\partial \eta_j}{\partial \xi_i} \frac{\partial \xi_i}{\partial \xi_k^0} \Xi_{kj}$$

$$= \left(\sum_{i,j} \frac{\partial \eta_j}{\partial \xi_i} \delta_{ij} \right) \frac{\partial \xi}{\partial \xi^0},$$

by (2.6). The definition of δ_{ij} shows that

$$\frac{\partial}{\partial t} \left(\frac{\partial \xi}{\partial \xi^0} \right) = \left(\sum_i \frac{\partial \eta_i}{\partial \xi_i} \right) \frac{\partial \xi}{\partial \xi^0}$$

$$= (\nabla_\xi \cdot \eta) \frac{\partial \xi}{\partial \xi^0}. \tag{2.9}$$

With (2.9) in hand, we can turn to the transport theorem. To evaluate the integral in (2.3), transform the domain $V(t)$ into $V(0)$, via (2.1). We find

$$\int_{V(t)} F(t, \xi) \, d\xi = \int_{V(0)} F(t, \xi \, (t, \xi^0)) \frac{\partial \xi}{\partial \xi^0} \, d\xi^0.$$

In this form, the domain of integration no longer depends on t, and if F is smooth, the derivative can be computed by differentiating under the integral sign. Using (2.9), we find

$$\frac{d}{dt} \int_{V(t)} F(t, \xi) \, d\xi = \int_{V(0)} [F_t + \nabla_\xi F \cdot \eta + F(\nabla_\xi \cdot \eta)] \frac{\partial \xi}{\partial \xi^0} \, d\xi^0$$

$$= \int_{V(t)} [F_t + \nabla_\xi \cdot (F\eta)] \, d\xi. \tag{2.10}$$

(2.10) is the transport theorem. To repeat, it is valid when F and $\xi(t, \xi^0)$ are smooth and the Jacobian is positive. We remind the reader that η is defined by (2.8).

3 Conservation of probability

Whatever else a real gas or liquid may be, it surely consists of a great many molecules interacting with each other because of mechanical, electrical, and other forces. We represent the molecules by point masses ("particles") and suppose for simplicity that all the molecules have the same mass. Then, the motion of the molecules is described by Newton's second law. This means that if there are N particles whose position at time t is described by N functions $q_1(t), \ldots, q_N(t)$, then these functions are solutions of the differential equations

$$m \frac{d^2 q_i}{dt^2} = Q_i(t, q_1, \ldots q_N), \, i = 1, \ldots, N. \tag{3.1}$$

Here Q_i is the force on the ith particle due to all the other particles. It is convenient to assume that the forces Q_i depend only on the variables shown in (3.1); in particular, we assume they do not depend on the velocities of the particles. However, Q_i may have terms in it that do not depend on the particles at all: for example, all the particles, in addition to interacting with each other, may be subject to the gravitational attraction of the earth.

Conditions for solutions of systems of ordinary differential equations like (3.1) to exist are well known. Of course, a solution may only exist for a small interval of time, or one may exist for all time. In the following, no hypotheses on the global existence of solutions of (3.1) are made, but we argue entirely on those values of t for which a solution does exist. Without further mention, we assume that we are talking about such values of t.

It is convenient to write (3.1) as a system of first order equations. For this purpose, define

$$\frac{dq_i}{dt} = s_i. \tag{3.2}$$

s_i is the velocity of the ith particle. It satisfies the equation

$$m \frac{ds_i}{dt} = Q_i(t, q_1, \ldots, q_N). \tag{3.3}$$

To determine a solution of (3.2–3) uniquely, we need the initial values of s_i and q_i. Write

$$q_i(0) = q_i^0$$

$$s_i(0) = s_i^0 \tag{3.4}$$

(3.2–3) can be written most simply in vector form. Formally, set $\xi = (q_1, \ldots, q_N, s_1, \ldots, s_N)$. Then equations (3.2–3) can be written simply

$$\frac{d\xi}{dt} = Q(t, \xi), \tag{3.5}$$

while (3.4) becomes

$$\xi(0) = \xi^0. \tag{3.6}$$

Notice that since each q_i and s_i is in R^3, ξ is an element of R^{6N}.

If N is at all large, it is clearly hopeless to expect to be able to measure the initial positions and velocities of all the particles initially. And even if this could somehow be done, no computer that we can imagine can solve the system (3.5) of $6N$ differential equations except, possibly, in extremely simple situations. Finally, even if all the initial data could be found and even if the differential equations (3.5) could be solved, it is hard to see how, given this information, it could be used to find out what we really want to know about the fluid the particles represent. Because of all these reasons, we pose the problem of determining the state of a large number of interacting particles in a different way.

A value of ξ in R^{6N} is called a *state* of the system of particles. We suppose that the state is initially determined, not precisely, but according to a certain probability distribution, and ask how the state is distributed at later times. To say that the state is initially distributed according to a probability distribution means that there is a real valued function F^0 defined on R^{6N} such that

$$F^0(\xi) \geq 0 \tag{3.7}$$

and

$$\int_{R^{6N}} F^0(\xi) \, d\xi = 1. \tag{3.8}$$

Any integrable function satisfying (3.7) and (3.8) is called a *probability distribution*. Let V be a domain in R^{6N}. The *probability* that the particles initially are in a state in V is defined to be

$$\int_V F^0(\xi) \, d\xi.$$

Clearly, (3.8) says only that the particles are initially in *some* state. We seek a probability distribution $F(t, \xi)$ satisfying

$$F(0, \xi) = F^0(\xi)$$

and one other condition that will be specified presently. Of course, $F(t, \xi)$ is to be interpreted as the probability distribution associated with the particles at time t which means, again, that the probability that the particles are in some state in the domain V at time t is

$$\int_V F(t, \xi) \, d\xi.$$

A solution of (3.5) is called a *trajectory*. Any trajectory is a function of $6N + 1$ real variables: t, and the coordinates of the initial point of the trajectory. Thus, a trajectory is a curve in R^{6N} described by a function

$$\xi = \xi(t, \xi^0) \tag{3.9}$$

with ξ^0 a fixed point in R^{6N}. Let $V(0)$ be a domain in R^{6N}. For each $\xi^0 \in V(0)$, we have a corresponding curve (3.9). When t is fixed, and ξ^0 varies over $V(0)$, $\xi(t, \xi^0)$ varies over a domain $V(t)$. Any family of domains $V(t)$ obtained in this way is said to *consist of trajectories*. Alternatively, one can say that a domain $V(t_0)$ consists of trajectories if every point in it is connected to a domain $V(0)$ by trajectories.

If $V(t)$ is any family of domains consisting of trajectories, we assume the probability that the particles are in a state in $V(t)$ at time t is independent of t. This hypothesis is called *conservation of probability*. Physically, what conservation of probability means is that particles do not wantonly appear or disappear, that a collection of particles moving along a certain trajectory does not suddenly change its mind and move along another. It is impossible to overemphasize the importance of the hypothesis of conservation of probability; it is the rock on which all continuum mechanics can ultimately be made to rest.

In terms of the probability distribution $F(t, \xi)$ describing the state of the particles at time t, the hypothesis of conservation of probability can be written in the form

$$\int_{V(t)} F(t, \xi) \, d\xi = \text{constant} \tag{3.10}$$

for any domains $V(t)$ consisting of trajectories.

There is a basic theorem on conservation of probability. To state it, we need a little more notation. Recall that ξ was defined originally as the vector

$$\xi = (q_1, \ldots, q_N, s_1, \ldots, s_N),$$

where q_i is a function describing the position, and s_i the velocity, of the ith particle. In accordance with the notational conventions established in § 1, we write

$$q_i = (x, y_i, z_i)$$

and

$$s_i = (u_i, v_i, w_i).$$

Define

$$\nabla_{q_i} = \left(\frac{\partial}{\partial x_i}, \frac{\partial}{\partial y_i}, \frac{\partial}{\partial z_i} \right)$$

and

$$\nabla_{s_i} = \left(\frac{\partial}{\partial u_i}, \frac{\partial}{\partial v_i}, \frac{\partial}{\partial w_i} \right).$$

THEOREM 3.1 *Consider a collection of particles moving according to Newton's law (3.1). Let $F^0(\xi)$ be any continuous probability distribution. Then there exists a unique probability distribution $F(t, \xi)$, continuous in all its variables and satisfying the two conditions:*

$$F(0, \xi) = F^0(\xi) \tag{3.11}$$

and conservation of probability. If $F^0(\xi)$ is differentiable, so is $F(t, \xi)$, and it satisfies the partial differential equation

$$F_t + \sum_{i=1}^{N} s_i \cdot \nabla_{q_i} F + \frac{1}{m} \sum_{s=1}^{N} Q_i \cdot \nabla_{s_i} F = 0. \tag{3.12}$$

Equation (3.12) is called *Liouville's equation*. For smooth initial distributions, Liouville's equation expresses the hypothesis of conservation of probability in a convenient way.

To break up the proof of the theorem into digestible parts, we state the following result separately.

LEMMA 3.2 *Consider a collection of particles moving according to Newton's law (3.1). Then, the Jacobian of any trajectory is given by the simple formula*

$$\frac{\partial \xi}{\partial \xi^0} \equiv 1.$$

Proof We use formula (2.9) for the derivative of the Jacobian. By definition

$$\xi = (q_1, \ldots, q_N, s_1, \ldots, s_N),\tag{3.13}$$

while

$$\eta = \xi_t.$$

By (3.2) and (3.8),

$$\eta = \left(s_1, \ldots, s_N, \frac{1}{m}Q_1, \ldots, \frac{1}{m}Q_N\right).$$

The q's and the s's are independent of each other. Moreover, by assumption, the Q's depend only on the q's. Therefore,

$$\nabla_\xi \cdot \eta = \sum \frac{\partial s_i}{\partial q_i} + \frac{1}{m}\sum \frac{\partial Q_i}{\partial s_i} = 0.$$

By (2.9), then, $\partial \xi / \partial \xi^0$ is constant. Since $\xi(0, \xi^0) = \xi^0$, $\partial \xi / \partial \xi^0$ is initially unity. Therefore,

$$\frac{\partial \xi}{\partial \xi^0} = 1$$

for all t. This proves the lemma.

If one is willing to assume the existence of a differentiable function F satisfying (3.10), Liouville's equation follows immediately from (3.10) and the lemma. Indeed, the transport theorem shows that

$$\frac{d}{dt}\int_{V(t)} F(t, \xi)\, d\xi = \int_{V(t)} [F_t + \nabla_\xi \cdot (F\eta)]\, d\xi.$$

By (3.10), this is zero for any $V(t)$ consisting of trajectories. For brevity, write $G = F_t + \nabla_\xi \cdot (F\eta)$. Because of the lemma, we have

$$\int_{V(0)} G(t, \xi(t, \xi^0))\, d\xi^0 = \int_{V(t)} G(t, \xi)\, d\xi = 0.$$

Since $V(0)$ is an arbitrary domain, this shows that G must be identically zero. Therefore,

$$0 = F_t + \nabla_\xi \cdot (F\eta)\tag{3.14}$$

$$= F_t + \sum_{i=1}^{N} \nabla_{q_i} \cdot (Fs_i) + \frac{1}{m}\sum_{i=1}^{N} \nabla_{s_i} \cdot (FQ_i).\tag{3.15}$$

Again using the facts that the q's and the s's are independent and that Q_i is independent of the s's, we derive Liouville's equation.

Since (3.15) is equivalent to Liouville's equation, and since (3.15) is sometimes the more convenient form, we also refer to (3.15) as Liouville's equation.

To prove the existence of F, we solve Liouville's equation formally. This will tell us what F should be, if it does exist. We then prove that this formal solution satisfies the conditions of the theorem. Now, any differentiable solution of Liouville's equation is constant on trajectories. Indeed, if $F(t, \xi)$ is a solution of Liouville's equation, then $F(t, \xi(t, \xi^0))$ is its value on a trajectory. But

$$\frac{\partial}{\partial t} F(t, \xi(t, \xi^0)) = F_t + \sum F_{\xi_i} \frac{\partial \xi_i}{\partial t} = F_t + (\nabla_\xi F) \cdot \eta.$$

On the other hand,

$$\nabla_\xi \cdot (F\eta) = \nabla_\xi F \cdot \eta + F(\nabla_\xi \cdot \eta)$$

and, because of the form of η, $(\nabla_\xi \cdot \eta) = 0$. Therefore,

$$\frac{\partial}{\partial t} F(t, \xi(t, \xi^0)) = F_t + \nabla_\xi \cdot (F\eta) = 0,$$

by (3.14). Thus, on trajectories, F is independent of t.

Let $F^0(\xi)$ be a given continuous function of ξ. We want to construct a function $F(t, \xi)$ that equals $F^0(\xi)$ when $t = 0$ and is constant on trajectories. Because of the lemma, the map from ξ^0 to ξ defined by

$$\xi = \xi(t, \xi^0) \tag{3.16}$$

is one-to-one. Therefore, (3.16) can be inverted to give ξ^0 as a function of ξ:

$$\xi^0 = \xi^0(t, \xi).$$

The function ξ^0, being the inverse of (3.16), has the property that

$$\xi^0(t, \xi(t, \xi^0)) \equiv \xi^0. \tag{3.17}$$

Define $F(t, \xi)$ by the formula

$$F(t, \xi) = F^0(\xi^0(t, \xi)). \tag{3.18}$$

We now prove that this is a continuous probability distribution satisfying (3.11) and conservation of probability.

First, F is surely a probability distribution, since F^0 is. For by (3.18), F is non-negative whenever F^0 has this property. Also,

$$\int_{R^{6N}} F(t, \xi)\, d\xi = \int_{R^{6N}} F^0(\xi^0(t, \xi))\, d\xi.$$

$$= \int_{R^{6N}} F^0(\xi^0)\, d\xi^0,$$

by the lemma. This integral is equal to unity since F^0 is a probability distribution. Therefore, F is a probability distribution.

The continuity of F is clear, since both $\xi^0(t, \xi)$ and $F^0(\xi)$ are continuous. Moreover,

$$F(0, \xi) = F^0(\xi^0(0, \xi))$$

$$= F^0(\xi)$$

since when $t = 0$ the map (3.12) is just the identity. Finally, F is constant on trajectories, since

$$F(t, \xi(t, \xi^0)) = F^0(\xi^0(t, \xi(t, \xi^0)))$$

$$= F^0(\xi^0),$$

by (3.17). This implies that F satisfies (3.10). For, making the transformation (3.16), the lemma shows that

$$\int_{V(t)} F(t, \xi)\, d\xi = \int_{V(0)} F(t, \xi(t, \xi^0))\, d\xi^0$$

$$= \int_{V(0)} F^0(\xi^0)\, d\xi^0,$$

and this is a constant. This shows that there is a continuous probability distribution satisfying (3.11) and conservation of probability.

To complete the proof of the theorem, we show that these conditions determine F uniquely. If V is any domain, we let $|V|$ denote its volume. Let $\xi^0 \in R^{6N}$, and let $V_n(0)$ be a sequence of domains shrinking down to ξ^0. Let $V_n(t)$ be the corresponding sequence of domains consisting of trajectories. By the lemma,

$$\int_{V_n(t)} F(t, \xi)\, d\xi = \int_{V_n(0)} F(t, \xi(t, \xi^0))\, d\xi^0.$$

To save writing, set $H(t, \xi^0) = F(t, \xi(t, \xi^0))$. The mean value theorem shows that for every n there is a point $\xi^n(t) \in V_n(0)$ such that

$$\int_{V_n(t)} F(t, \xi)\, d\xi = H(t, \xi^n(t))\, |V_n(0)|.$$

If F satisfies conservation of probability, this integral must be independent of t. Since $|V_n(0)|$ is clearly independent of t, this shows that $H(t, \xi^n(t))$ is independent of t. Since $\xi^n(t) \in V_n(0)$, while $V_n(0)$ shrinks down to ξ^0, we must have $\xi^n(t) \to \xi^0$ as $n \to \infty$. Since F is continuous, H must be continuous. Therefore,

$$H(t, \xi^0) = \lim_{n \to \infty} H(t, \xi^n(t)),$$

and this shows that since $H(t, \xi^n(t))$ is independent of t, $H(t, \xi^0)$ is also independent of t for any $\xi^0 \in R^{6N}$. Going back to the definition of H, we see that this is the same as saying that F is constant on trajectories. Finally, if F is constant on trajectories and satisfies (3.11),

$$F(t, \xi(t, \xi^0)) = F(0, \xi(0, \xi^0))$$

$$= F(0, \xi^0)$$

$$= F^0(\xi^0).$$

This is the same as (3.18) and shows that F is uniquely determined.

Exercises

3.1 Let $s_i = dq_i/dt$. Suppose we have a collection of particles of mass m whose motion is described by the differential equations

$$m\frac{d^2q_i}{dt^2} = Q_i(t, q_1, \ldots, q_N, s_1, \ldots, s_N),$$

where the forces Q_i depend on the s's and not, as in the text, merely on the q's. Derive the equation corresponding to Liouville's equation that expresses conservation of probability in this case.

3.2 Consider a collection of N particles, the first N_1 of which have mass m_1, the rest having mass m_2. The motion of the particles is then described by equations of the form

$$m_1\frac{d^2q_i}{dt^2} = Q_i(t, q_1, \ldots, q_N), i = 1, \ldots, N_1,$$

$$m_2\frac{d^2q_i}{dt^2} = Q_i(t, q_1, \ldots, q_N), i = N_1 + 1, \ldots, N.$$

In this case, express conservation of probability in a form resembling Liouville's equation.

4 The conservation equations. Definition of a fluid

In any real fluid, all molecules having the same mass, chemical constituency, etc., are indistinguishable. In our considerations, we assume for simplicity that *all* the molecules are indistinguishable. This manifests itself in the forces Q_i occurring in (3.1): they have a high degree of symmetry. The usual hypothesis in statistical mechanics is that the forces have the form

$$Q_i(t, q_1, \ldots, q_N) = \sum_{j=1}^{N} Q_{ij}(t, |q_i - q_j|).$$

For our purposes, we don't need quite this much—that Q_t is a symmetric function of the variables $q_1, \ldots, q_N$ will do. In this section, we begin with the probability distribution $F(t, \xi)$ and average out all but one of the variables $q_1, \ldots, q_N$ to derive certain equations for the distinguished particle. It is clear from Liouville's equation that the symmetry of the Q's is inherited by F. Therefore, it does not matter which of the N variables q_t we choose to distinguish; the first will do as well as any. To simplify the notation, we write q for q_1, and ζ for $(q_2, \ldots, q_N, s_1, s_2, \ldots, s_N)$. In this notation, we write $F(t, q, \zeta)$ instead of $F(t, \xi)$. Clearly, $q \in R^3$, and $\zeta \in R^{6N-3}$.

In a while, we shall define a fluid as a system of particles satisfying conservation of probability and one other condition. Before doing so, we define certain functions that later on will be identified with functions that arise intuitively in any naive consideration of a continuous medium. The functions that we have in mind are called in probability theory the *moments* of the distribution F, a moment being any integral of the form

$$\int_{R^{6N-3}} P(s_1, \ldots, s_N) F(t, q, \zeta) \, d\zeta,$$

where P is a polynomial.

To be specific, consider a system of particles of mass m satisfying (3.1). Let $F(t, q, \zeta)$ be a probability distribution satisfying Liouville's equation (3.12). Define a function ϱ by the formula

$$\varrho(t, q) = \int_{R^{6N-3}} m F(t, q, \zeta) \, d\zeta. \tag{4.1}$$

For any domain V,

$$\int_V F(t, q, \zeta) \, d\zeta \, dq = \int_V F(t, \zeta) \, d\zeta$$

si a probability, and as such it is dimensionless. m, of course, has the dimensions of mass. It follows that (4.1) has the dimensions: mass/volume. A quantity with such dimensions is usually called a density; accordingly, we call (4.1) the *density of the system of particles* at the point (t, q).

F is a function of $t, q, q_2, \ldots, q_N, s_1, s_2, \ldots, s_N$. Define a function s by the formula

$$s(t, q) = \frac{1}{\varrho(t, q)} \int_{R^{6N-3}} m s_1 F(t, q, \zeta) \, d\zeta. \tag{4.2}$$

s_1 is a 3-vector, having the dimensions of a velocity. It is easy to see, then, that s has the dimensions of a velocity, and we call s the *velocity of the system of particles* at the point (t, q).

There is an equation relating the functions ϱ and s. To derive it, we remind the reader of the *divergence theorem*. Consider a domain V in R^m. Let ζ be a generic point in R^m, with coordinates $(\zeta_1, \ldots, \zeta_m)$ in some rectangular coordinate system. We write

$$\nabla_\zeta = \left(\frac{\partial}{\partial \zeta_1}, \ldots, \frac{\partial}{\partial \zeta_m} \right).$$

Let f be a smooth function, with values in R^m, defined on the closure of a domain $V \subset R^m$. Finally, let n be the outward normal to the boundary of V, and let dA be the area element on the boundary of V. Then, the divergence theorem says that

$$\int_V \nabla_\zeta \cdot f \, d\zeta = \int_{\partial V} n \cdot f \, dA. \tag{4.3}$$

Consider now Liouville's equation in the form (3.15). We have been writing $\zeta = (q_2, \ldots, q_N, s_1, s_2, \ldots, s_N)$. Therefore,

$$\nabla_\zeta = (\nabla_{q_2}, \ldots, \nabla_{q_N}, \nabla_{s_1}, \nabla_{s_2}, \ldots, \nabla_{s_N}).$$

Write

$$a = (ms_2, \ldots, ms_N, Q_1, Q_2, \ldots, Q_N). \tag{4.4}$$

Clearly $a \in R^{6N-3}$. Recalling that we have written q instead of q_1, we see that (3.15) can be written in the form

$$mF_t + m\nabla_q \cdot (s_1 F) + \nabla_\zeta \cdot (aF) = 0. \tag{4.5}$$

Let B_r be the ball of radius r centered at the origin in R^{6N-3}. Integrate (4.5) over B_r. Since t, q, and ζ are independent variables, we have

$$\frac{\partial}{\partial t} \int_{B_r} mF \, d\zeta + \nabla_q \cdot \int_{B_r} s_1 F \, d\zeta = -\int_{B_r} \nabla_\zeta \cdot (aF) \, d\zeta$$

$$= -\int_{\partial B_r} n \cdot (aF) \, dA,$$

by the divergence theorem (4.3). As $r \to \infty$, the first two terms here go to ϱ_t and $\nabla_q \cdot (\varrho s)$, by (4.1) and (4.2). Therefore, we have

$$\varrho_t + \nabla_q \cdot (\varrho s) = -\lim_{r \to \infty} \int_{\partial B_r} n \cdot (aF) \, dA. \tag{4.6}$$

We saw in §3 that any solution of Liouville's equation is constant on trajectories. Therefore, if the initial distribution $F^0(\xi)$ goes to zero as $|\xi| \to \infty$, the same is true of $F(t, \xi)$. Thus, the right side of (4.6) is zero, and we have

$$\varrho_t + \nabla_q \cdot (\varrho s) = 0. \tag{4.7}$$

This is the equation relating ϱ and s. It is usually called the *continuity equation*. We shall see later on that it has a physical interpretation as *conservation of mass*. Great care has not been taken here with the conditions F^0 must satisfy as $|\xi| \to \infty$; all we have really proved is that if F goes to zero fast enough at infinity, then (4.7) holds. We state precise conditions for (4.7) to hold in the exercises below.

Now, let

$$s_1 - s(t, q) = S(t, q, s_1). \tag{4.8}$$

Clearly, the definition (4.2) of $s(t, q)$ implies

$$\int_{R^{6N-3}} mSF \, d\zeta = 0. \tag{4.9}$$

S is a 3-vector. In the coordinate system given in R^3, let its components be (S_1, S_2, S_3). We define a matrix $\tau(t, q)$ as follows. The element in the ith row and the jth column of $\tau(t, q)$ is

$$\tau_{ij}(t, q) = \int_{R^{6N-3}} mS_i S_j F \, d\zeta. \tag{4.10}$$

By the same method as before, we can show that each τ_{ij} has the dimensions of a pressure: force/area. The matrix τ has a name; it is called the *stress tensor*[1] *of the system of particles*.

There is an equation like (4.7) relating ϱ, s, and τ. To derive it, it is convenient to change our earlier notation and to denote the components of q, s, and s_1 by (q^1, q^2, q^3), (s^1, s^2, s^3), and (s_1^1, s_1^2, s_1^3), respectively. Multiply Liouville's equation in the form (4.5) by s_1^i, and integrate over the ball B_r of radius r in R^{6N-3}. There results

$$\int_{B_r} ms_1^i F_t \, d\zeta + \sum_j \int_{B_i} ms_1^i \frac{\partial}{\partial q^j} (s_1^j F) \, d\zeta + \int_{B_r} s_1^i \nabla_\zeta \cdot (aF) \, d\zeta = 0. \tag{4.11}$$

We examine each of the terms in this equation separately. Since m and s_1^i are independent of t, we have, first of all,

$$\int_{B_r} ms_1^i F_t \, d\zeta = \frac{\partial}{\partial t} \int_{B_r} ms_1^i F \, d\zeta$$

$$\to \frac{\partial}{\partial t} (\varrho s^i) \tag{4.12}$$

as $r \to \infty$, by (4.2).

---

[1] Until chapter seven, when the tensorial character of τ becomes important, we use the word tensor as a synonym for matrix.

Next, since m and s_1^i are independent of q^j,

$$\sum_j \int_{B_r} m s_1^i \frac{\partial}{\partial q^j} (s_1^j F) \, d\zeta = \sum_j \frac{\partial}{\partial q^j} \int_{B_r} m s_1^i s_1^j F \, d\zeta$$

$$= \sum_j \frac{\partial}{\partial q^j} \int_{B_r} m (s^i + S^i)(s^j + S^j) F \, d\zeta$$

$$= \sum_j \frac{\partial}{\partial q^j} \int_{B_r} m (s^i s^j + S^i S^j) F \, d\zeta,$$

by (4.9). Therefore,

$$\sum_j \int_{B_r} m s_1^i \frac{\partial}{\partial q^j} (s_1^j F) \, d\zeta = \sum_j \frac{\partial}{\partial q^j} \left(s^i s^j \int_{B_r} m F \, d\zeta + \int_{B_r} m S^i S^j F \, d\zeta \right)$$

$$\rightarrow \sum_j \frac{\partial}{\partial q^j} (\varrho s^i s^j + \tau_{ij}) \tag{4.13}$$

as $r \rightarrow \infty$, by (4.1) and (4.10).

Finally, we turn to the last term in (4.11). ∇_ζ has the form

$$\nabla_\zeta = (\nabla_{q_2}, \ldots, \nabla_{q_N}, \nabla_{s_1}, \nabla_{s_2}, \ldots, \nabla_{s_N}).$$

Except for the component ∇_{s_1}, each entry in ∇_ζ commutes with multiplication by s_1. Therefore, again with a single exception, the quantity s_1^i appearing in the last term of (4.11) can be brought inside the divergence symbol. Assuming that F goes to zero fast enough, these terms can then be shown to go to zero as $r \rightarrow \infty$, by the divergence theorem. There remains the term

$$\int_{B_r} s_1^i \nabla_{s_1} \cdot (Q_1 F) \, d\zeta. \tag{4.14}$$

(Recall the form, (4.4), of a.) Denote the components of Q_1 by (Q_1^1, Q_1^2, Q_1^3). (4.14) equals

$$\sum_j \int_{B_r} s_1^i \frac{\partial}{\partial s_1^j} (Q_1^j F) \, d\zeta = \sum_j \int_{B_r} \frac{\partial}{\partial s_1^j} (s_1^i Q_1^j F) \, d\zeta - \sum_j \int_{B_r} Q_1^j F \frac{\partial s_1^i}{\partial s_1^j} \, d\zeta.$$

The first term here vanishes as $r \rightarrow \infty$. As for the second, we have

$$\frac{\partial s_1^i}{\partial s_1^i} = \delta_{ij}.$$

As $r \rightarrow \infty$, then, all that is left of the last term in (4.11) is

$$-\int_{R^{6N-3}} Q_1^i F \, d\zeta.$$

Using this result, as well as (4.12) and (4.13), in (4.11), we find

$$\frac{\partial}{\partial t}(\varrho s^i) + \sum_j \frac{\partial}{\partial q^j}(\varrho s^i s^j + \tau_{ij}) = \int_{R^{6N-3}} Q_1^i F \, d\zeta.$$

(4.7) can be used to eliminate the derivative of ϱ with respect to t. The result is

$$\frac{\partial s^i}{\partial t} + \sum_j s^j \frac{\partial s^i}{\partial q^j} + \frac{1}{\varrho} \sum_j \frac{\partial}{\partial q^j} \tau_{ij} = \frac{1}{\varrho} \int_{R^{6N-3}} Q_1^i F \, d\zeta. \qquad (4.15)$$

Write $\nabla_q \cdot \tau$ for the vector whose ith component is $\sum_j (\partial/\partial q^j)\tau_{ij}$. Then, (4.15) becomes

$$s_t + (s \cdot \nabla_q)\, s + \frac{1}{\varrho} \nabla_q \cdot \tau = \frac{1}{\varrho} \int_{R^{6N-3}} Q_1 F \, d\zeta. \qquad (4.16)$$

(4.16), like (4.17), has a physical interpretation as a conservation equation; we shall see later on that it represents *conservation of momentum*.

Before turning to this matter, let us see where we are. We defined three quantities, ϱ, s, and τ in terms of the first few moments of F. We found relations (4.7) and (4.16) connecting these quantities. For some purposes, it is necessary to consider higher order moments of F. Again, there are relations connecting these moments, and all of these relations have significance ce conservation equations. For example, if Liouville's equation (4.5) is multiplied by $s_1^i s_1^j$ and integrated over R^{6N-3}, the equation obtained in this way can be interpreted as representing *conservation of energy*.[1]

But let us see what is happening as we derive these conservation equations. We consider ϱ, s, τ, etc. as unknown functions. ϱ is a scalar, and s is a 3-vector. Therefore, ϱ and s are determined by *four* unknown scalar quantities. To connect these unknowns, we have the single conservation equation (4.7). Consideration of τ adds six more scalar unknowns. (τ is a 3×3 matrix, but its definition, (4.10), shows that it is symmetric.) On the other hand, forgetting about the term on the right for a moment, (4.16) represents three scalar equations connecting ϱ, s, and τ. Thus, (4.7) is one equation connecting four unknowns. (4.7) and (4.16) together are four scalar equations connecting ten unknowns. If we were to consider the energy equation, we should obtain six more scalar equations, but more new unknowns yet would be added.

It turns out that the moments of F are the significant quantities in fluid mechanics; they are much more significant physically than F itself. But it is clear that we can never hope to determine the moments from the conserva-

---

[1] For a derivation of this equation and a discussion of it (in a slightly different context) see Harold Grad, *Principles of the kinetic theory of gases*. In the Encyclopedia of Physics, vol. XII, S. Flügge, ed. Springer-Verlag, Berlin, 1958.

tion equations unless we are willing to consider infinitely many such equations and infinitely many of the moments, since the number of unknown moments always far exceeds the number of conservation equations.

This difficulty will be eliminated by edict. To see how this is done, we need a few more definitions. The quantity

$$p = \tfrac{1}{3} \operatorname{tr} \tau = \tfrac{1}{3} \sum_{i=1}^{3} \tau_{ii} \qquad (4.17)$$

is called the *pressure of the system of particles*. (Recall that we have seen that each τ_{ij} has the dimensions of a pressure.) Let I be the identity matrix, and write

$$\sigma = \tau - pI. \qquad (4.18)$$

We call σ the *reduced stress tensor of the system*. Finally, let D be the matrix whose ijth entry is

$$\frac{1}{2} \left(\frac{\partial s^i}{\partial q^j} + \frac{\partial s^j}{\partial q^i} \right). \qquad (4.19)$$

D is called the *deformation tensor of the system*.

With these definitions in hand, we can define a fluid. A system of N particles satisfying conservation of probability is called a *fluid*[1] if the reduced stress tensor σ is a function only of the deformation tensor[2] D. This definition of a fluid is usually called *the Stokes hypothesis*.

Let us see what the connection between the number of moments and the number of conservation equations is for fluids. We still have the four unknowns ϱ and the components of s. The pressure p is another unknown scalar. But the components of σ are given in terms of the components of s since σ is a (known) function of D, and D depends only on s. Thus, we have *five* scalar unknowns in all: ϱ, p, and the components of s. On the other hand, we have *four* scalar conservation equations, (4.7) and the vectorial equations (4.16). To make the number of equations equal to the number of unknowns, a relation between ϱ and p is usually added. This relation is called the *equation of state of the fluid*. Given an equation of state, the number of unknowns and the number of equations from which they are to be determined is the same, and there is a chance at least of the equations determining the unknowns uniquely.

We now write down the relations we have so far found between the moments in the original notation of § 1. q is an independent variable in R^3. We

---

[1] The term is meant to include both gases and liquids.

[2] Certain technical requirements are also imposed on the function relating ϱ and D. We discuss these in chapter seven.

suppose that a rectangular coordinate system has been introduced in R^3. In this coordinate system, we write

$$q = (x, y, z).$$

As in § 1, we write ∇ instead of ∇_q. Notice that since σ and τ are related by (4.18), the definition of $\nabla \cdot \tau$ gives

$$\nabla \cdot \tau = \nabla p + \nabla \cdot \sigma.$$

Therefore, equations (4.7) and (4.16) have the form

$$\varrho_t + \nabla \cdot (\varrho s) = 0, \tag{4.20}$$

$$s_t + (s \cdot \nabla) s + \frac{1}{\varrho} \nabla \cdot \sigma = -\frac{1}{\varrho} \nabla p + \frac{1}{\varrho} \int_{R^{6N-3}} Q_1 F \, d\zeta. \tag{4.21}$$

In addition the Stokes hypothesis is

$$\sigma = f(D). \tag{4.22}$$

If, in the given coordinate system,

$$s = (u, v, w),$$

then D has the form

$$D = \begin{pmatrix} u_x & \frac{1}{2}(u_y + v_x) & \frac{1}{2}(u_z + w_x) \\ \frac{1}{2}(u_y + v_x) & v_y & \frac{1}{2}(v_z + w_y) \\ \frac{1}{2}(u_z + w_x) & \frac{1}{2}(v_z + w_y) & w_z \end{pmatrix}. \tag{4.23}$$

In addition, there is an equation of state. In all our later considerations, we take the simplest such equation

$$\varrho = \text{constant}. \tag{4.24}$$

Fluids in which (4.24) is satisfied are called *incompressible*. Since F is a probability distribution (and, therefore, never negative), (4.1) shows that the constant in (4.24) must be positive. When the fluid is incompressible, (4.20) takes the particularly simple form

$$\nabla \cdot s = 0. \tag{4.25}$$

Finally, we attempt to say what all this has to do with fluids as such things are normally conceived. The definition (4.1) of ϱ shows that it is a kind of average density of the particles under consideration. Similar interpretations apply to s, as an average velocity, and to p, as an average pressure. We think of a fluid as a continuous medium, at each point having as its density the average density of the particles, as its velocity the average velocity of the particles, and as its pressure the average pressure of the particles. The conservation equations and the equation of state then relate all these quantities. We shall derive these equations again in § 7, assuming a little more physical insight into real fluids.

Exercises

4.1 Let $F^0(\xi)$ be identically zero outside a bounded set of values of ξ. Interpret this condition physically. Using the fact that $F(t, \xi)$ is constant on trajectories, show that $F(t, \xi)$ is also zero outside a bounded set, and prove carefully that in this case equations (4.7) and (4.16) are valid.

4.2 Using the fact that $F(t, \xi)$ is constant on trajectories, show that

$$\int_{R^{6N}} (1 + |\xi| + |\xi|^2)\, F(t, \xi)\, d\xi < \infty$$

if

$$\int_{R^{6N}} (1 + |\xi| + |\xi|^2)\, F^0(\xi)\, d\xi < \infty. \tag{4.26}$$

Assuming (4.26), then show carefully that (4.7) and (4.16) are valid.

5 The Stokes hypothesis

We wrote down the Stokes hypothesis (4.22) baldly, without any justification whatever. An heuristic, physical argument for it will be presented in chapter seven. Before getting to that, it is perfectly reasonable to pose the following purely mathematical question: for what initial distributions $F^0(\xi)$ and what forces Q_i is the Stokes hypothesis satsified? This question makes perfect sense since, when F^0 and Q_i are given, the Liouville equation can be solved for F, as we saw in theorem 3. Once F is known, s, D, and σ can be computed from (4.1), (4.2), (4.10), (4.18), and (4.23). Having computed these quantities, one can ask if the computed σ is a function of the computed D, as the Stokes hypothesis requires. The answer to the question is this: in *no* known nontrivial situation is σ a function of D. Thus, as far as is known, the Stokes hypothesis is *never* satisfied!

We should be clear about this. There is no question that in *real* fluids, under ordinary conditions, the Stokes hypothesis is satisfied; all experiments are in agreement about that. There is also no question about the validity of the conservation equations. The only question is the mathematical one, does the Stokes hypothesis follow from Liouville's equation? Or better, *when* does the Stokes hypothesis follow, and when does it not?

There is an equation related to the Liouville equation from which the

Stokes hypothesis can be formally derived. This equation is called the *Boltzmann equation*. The Boltzmann equation can be formally derived from the Liouville equation, via certain limiting processes and using certain "physically reasonable" hypotheses. The fact is, however, that the Boltzmann equation is *inconsistent* with the Liouville equation. For this reason, the Boltzmann equation has been a source of controversy since its creation in the nineteenth century. Its inconsistency with Liouville's equation notwithstanding, physicists will not give up Boltzmann's equation, and for a very good reason. Certain easily verifiable physical facts follow from Boltzmann's equation that do *not* follow from Liouville's equation. These facts can be proved not to follow from Liouville's equation, and, in fact, it is this proof that shows the two equations to be inconsistent[1].

If one is willing to overlook the discrepancies between the Boltzmann and the Liouville equations and begin with the Boltzmann equation as fundamental, both the Stokes hypothesis and the conservation equations can be derived by expanding the solution in a series of powers of a certain parameter. But there is still work for the mathematician to do. It is not known if the expansion *ever* converges, or, indeed, what sense can be made of it at all as anything other than a purely formal technique for reaching conclusions already decided upon on physical grounds. This expansion—due, by the way, to Hilbert—can be found in the work by Grad already cited.

A question like the one we asked about the Stokes hypothesis can also be asked about the density. An incompressible fluid was defined as one for which ϱ is constant. But is this ever true? If F^0 and Q_i are given, F can be found, and then ϱ can be computed from (4.1). For what F^0 and Q_i is the computed ϱ constant? Again, nothing is known about this. The question can be weakened to: when is ϱ *approximately* constant? Here, *something* is known. In the simplest of all cases, when all $Q_i = 0$, and when the fluid is contained in a rectangular box, then $\varrho\,(t, q)$ approaches a constant as $t \to \infty$. This is proved in the next section.

Exercise

5.1 As a start in interpreting ϱ physically as what is normally thought of as a density, prove the following result. Let V be a domain in R^3 into which no particles can ever enter. (This means that the probability is zero that a particle is in V.) Show that if $q \in V$, then $\varrho\,(t,q) = 0$.

---

[1] For a justification of the Boltzmann equation, see Harold Grad, *op. cit.* Grad also attempts a reconciliation of the two equations. For a discussion of the Boltzmann equation in its philosophical setting, see Max Born, *op. cit.*

6 Boundary conditions. A theorem of Grad

In many problems that occur in practice, whether for fluids or simply for systems of particles, one is not interested—even if it were possible—in simply solving Newton's equations, or Liouville's equation, or even the conservation equations, for given initial data with no other conditions. To take the case of a fluid, it is usually contained initially between certain boundaries, and it must remain so. Milk is in a bottle; the ocean is bounded by the sea floor; even the atmosphere does not penetrate the surface of the earth. We wish to express such conditions as side conditions imposed on the solution of Liouville's equation.

Suppose we have a system of particles satisfying Newton's equations (3.1), with the constraint that they originate and must stay in a domain V. If q_i^0, the initial position of the ith particle, is in V, then the particle will move according to (3.1) until it hits the boundary of V. Unless its path is tangent to V, something new must happen at this point if the particle is to remain in V. Various hypotheses are possible. The simplest is that the particle simply reflects off ∂V, exactly as if the boundary were a mirror. When this happens, the phenomenon is called *specular reflection*. Let n be the outer and n' the inner unit normal to ∂V at the point of incidence of the particle. Denote the velocity of the particle when it strikes the boundary by s_i. The velocity of the particle when it leaves the boundary must be different from s_i if the particle is to remain in V. Let s_i' be the velocity of the particle when it leaves ∂V. The condition of specular reflection is that s_i' lies in the plane spanned by n and s_i, that the length of s' is the same as the length of s_i, and that the angle between s_i' and n' is the same as the angle between s_i and n. A little vector algebra that this entails

$$s_i' = s_i - 2\,(n \cdot s_i)\,n. \tag{6.1}$$

After striking the boundary, the particle leaves from the same point it struck, but at the new velocity (6.1). From this point until it next strikes the boundary, the particle's motion is again determined by Newton's equations (3.1).

We must interpret the condition of specular reflection in terms of the probability distribution F. As we remarked, if the particles are indistinguishable, F is a symmetric function of the particles, and to interpret specular reflection in terms of F, we may as well interpret it for the first particle—its behavior with respect to the other particles is then the same. F is a function of t and of

$$\xi = (q_1, \ldots, q_N, s_1, \ldots, s_N).$$

We must consider the variables q_1 and s_1 separately from the rest. Denote the vectors $q_2, \ldots, q_N$ collectively by q_* and the vectors $s_2, \ldots, s_N$ by s_*. We write

$$F(t, \xi) = F(t, q_1, q_*, s_1, s_*).$$

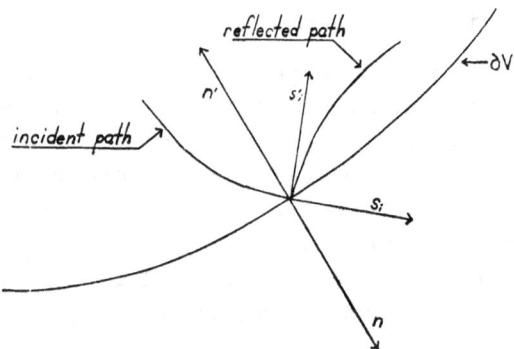

Figure 1

As we saw in § 3, F is constant on trajectories. When $q_1 \in \partial V$, the only way F can be constant on trajectories is for F to take on the same values at (q_1, q_*, s_1', s_*) and at (q_1, q_*, s_1, s_*). In view of (6.1), this means that

$$F(t, q_1, q_*, s_1, s_*) = F(t, q_1, q_*, s_1 - 2\,(n \cdot s_1)\,n, s_*),$$

$$\text{whenever} \quad q \in \partial V. \quad (6.2)$$

F is still constant on trajectories. What (6.2) says is, that although trajectories may have corners when one of the particles lies on the boundary of V, F remains constant. Note that if F satisfies (6.2), its initial value, F^0, must surely do so.

We now state one of the few known results about the moments of F for a fluid. A system of particles is called a *Knudsen gas* if all the forces Q_i occurring in (3.1) are zero. In a Knudsen gas, there are no external forces (such as gravity), and the particles themselves never interact with each other; however, the particles may interact with the walls of the container. Any very rare gas under the influence of no external forces approximates a Knudsen gas since the gas being rare means there are few molecules in the container and, therefore, that they cannot interact very often.

For a Knudsen gas, Liouville's equation can be greatly simplified. To begin, notice that for a Knudsen gas, Liouville's equation has the form

$$F_t + \sum_{i=1}^{N} s_i \cdot \nabla_{q_i} F = 0.$$

Write

$$G(t, q_1, s_1) = \int_{R^{6N-6}} F(t, q_1, q_*, s_1, s_*)\, dq_*\, ds_*. \quad (6.3)$$

Note that since F is a probability distribution on R^{6N}, G is a probability distribution on R^5. As in § 4, the divergence theorem can be used to show that if F goes to zero fast enough at infinity, then G satisfies

$$G_t + s_1 \cdot \nabla_{q_1} G = 0. \tag{6.4}$$

Thus G, which depends only on the first particle and is independent of the others, satisfies a one-particle Liouville equation. Also,

$$\varrho\,(t, q_1) = \int_{R^{6N-3}} mF\,(t, q_1, q_*, s_1, s_*)\, dq_*\, ds_*\, ds_1$$

$$= \int_{R^3} mG\,(t, q_1, s_1)\, ds_1. \tag{6.5}$$

Finally, if the particles reflect specularly off the boundary of a domain V, then (6.2) shows that

$$G\,(t, q_1, s_1) = G\,(t, q_1, s_1 - 2\,(n \cdot s_1)\,n) \quad \text{whenever} \quad q_1 \in \partial V. \tag{6.6}$$

In the following theorem, we denote the initial value of G by G^0. Of course,

$$G^0\,(q_1, s_1) = \int_{R^{6N-6}} F^0\,(q_1, q_*, s_1, s_*)\, dq_*\, ds_*.$$

THEOREM 6.1   *Let V be a rectangular parallelopiped in R^3. Consider a Knudsen gas confined to V with specular reflection from the walls of V. If the initial distribution, $G^0\,(q_1, s_1)$, is uniformly continuous in s_1 and satisfies*

$$G^0\,(q_1, s_1) \leqslant H(s_1), \tag{6.7}$$

where

$$\int_{R^3} H(s_1)\, ds_1 < \infty, \tag{6.8}$$

then the density of the system approaches a constant as t goes to infinity.

This attractive theorem is due to H. Grad.[1] It says that as times goes on, the particles distribute themselves more and more nearly uniformly throughout V, no matter how they were distributed initially. To see this, notice that according to the definitions of sections 3 and 4, the probability that some particle lies in a subdomain V_1 of V is just

$$\frac{1}{m} \int_{V_1} \varrho\,(t, q)\, dq.$$

Since ϱ goes to a constant, the probability that a particle lies in V_1 is ultimately proportional simply to the volume of V_1. This is just another way of saying that the distribution of particles is uniform.

---

[1] The many faces of entropy. Communications on pure and applied math., 14 (1961) 323–354.

To prove the theorem, let V be the parallelopiped

$$0 < x < a, \quad 0 < y < b, \quad 0 < z < c.$$

Since all the particles are initially in V, and since they reflect off the walls of V, no particle is ever outside V. As a consequence, F—and, therefore, G—is zero if q_1 is not in V. Suppose for a moment that G were known in V. We begin by redefining G outside V. For this purpose, let W be the doubled parallelopiped

$$0 < x < 2a, \quad 0 < y < 2b, \quad 0 < z < 2c.$$

In V, we define a function Γ as identically equal to G:

$$\Gamma(t, q_1, s_1) = G(t, q_1, s_1) \quad \text{for} \quad q_1 \in V. \tag{6.9}$$

We define Γ in W by reflection. This means that if $q_1 = (x, y, z)$, and if we write $G(t, x, y, z, s_1)$ for $G(t, q_1, s_1)$, then

$$\Gamma(t, a + x, y, z, s_1) = \Gamma(t, a - x, y, z, s_1),$$

with similar formulas for the other variables.

The function Γ thus defines satisfies periodic boundary conditions in W—that is, the value of Γ at any point on one face of W is the same as its value at the corresponding point on the opposite face. Outside W, we define Γ so that it is periodic:

$$\Gamma(t, x + 2la, y + 2mb, z + 2nc, s_1) = \Gamma(t, x, y, z, s_1)$$

for any integers l, m, and n.

Γ is now a function defined for all $q \in R^3$. Moreover, it has the same continuity properties as G, and, for $q \in V$, G can be recovered from Γ by (6.9). Finally, Γ also satisfies the one-particle Liouville equation

$$\Gamma_t + s_1 \cdot \nabla_{q_1} \Gamma = 0, \tag{6.10}$$

since G does.

$\Gamma(0, q_1, s_1) = \Gamma^0(q_1, s_1)$ is known for all q_1 and s_1, since $G^0(q_1, s_1)$ is given in V, and $\Gamma^0(q_1, s_1)$ can be found by the construction just outlined. Now, the solution of (6.10) with Γ given initially is easily found by the method used in the proof of theorem 3.1. The result is

$$\Gamma(t, q_1, s_1) = \Gamma^0(q_1 - s_1 t, s_1). \tag{6.11}$$

This is easily verified by substituting the function Γ defined by (6.11) into (6.10) and noticing that it has the right initial value.

With (6.9) and (6.11) in hand, we can complete the proof of Grad's theorem. We have

$$\varrho(t, q_1) = \int_{R^3} m G(t, q_1, s_1) \, ds_1 = \int_{R^3} m \Gamma^0(q_1 - s_1 t, s_1) \, ds_1$$

whenever $q_1 \in V$. If $\bar{q}_1$ is another point in V, we have

$$\varrho(t, \bar{q}_1) = \int_{R^3} m\Gamma^0(\bar{q}_1 - s_1 t, s_1) \, ds_1$$

$$= \int_{R^3} m\Gamma^0\left(q_1 - s_1 t, s_1 + \frac{\bar{q}_1 - q_1}{t}\right) ds_1,$$

where we have replaced s_1 by $s_1 + (\bar{q}_1 - q_1)/t$. Therefore,

$$|\varrho(t, q_1) - \varrho(t, \bar{q}_1)|$$

$$\leqslant \int_{R^3} m\left|\Gamma^0(q_1 - s_1 t, s_1) - \Gamma^0\left(q_1 - s_1 t, s_1 + \frac{\bar{q}_1 - q_1}{t}\right)\right| ds_1$$

$$\leqslant \int_{|s_1| \leqslant r} m\left|\Gamma^0(q_1 - s_1 t, s_1) - \Gamma^0\left(q_1 - s_1 t, s_1 + \frac{\bar{q}_1 - q_1}{t}\right)\right| ds_1$$

$$+ 2\int_{|s_1| > r} mH(s_1) \, ds_1, \tag{6.11}$$

by (6.7).

Take $\varepsilon > 0$. Because of (6.8), the second term in (6.11) can be made less than $\varepsilon/2$ by choosing r large enough. With r fixed, the first term in (6.11) can be made less than $\varepsilon/2$ by choosing t large enough, since Γ^0 is uniformly continuous in s_1. This shows that if t is large enough, then

$$|\varrho(t, q_1) - \varrho(t, \bar{q}_1)| < \varepsilon$$

for any $\varepsilon > 0$, which is what we wanted to prove.

As we pointed out earlier, the theorem says that for a Knudsen gas in a rectangular box, the particles tend to be distributed more and more randomly as time goes on. Presumably, if the system is not a Knudsen gas and the particles are allowed to interact, this tendency toward randomness can only be enhanced, but such a result has never been proved.

The proof above depends heavily on the fact that V is a rectangular box. Of course, no physical box is ever perfectly rectangular and—again presumably—if the sides of the box are not perfectly flat, the tendency to randomness is again enhanced. It would be most interesting to prove such a theorem, but the proof must be difficult, for the theorem is *false* if V is a sphere[1]! The fact that the theorem is false for some domains unfortunately deprives it of physical significance. Although one can argue that no real

---

[1] This was first pointed out to me by R. R. Welland.

container is ever perfectly spherical and, therefore, that the sphere is an anomaly, it can also be argued with equal force that no real container is ever a parallelopiped either. Perhaps it is the box that is an anomaly!

Exercise

6.1 Show that theorem 6 is false if V is either a sphere or a circular cylinder. (*Hint*: Use exercise 5.1.)

7 Fluid mechanical derivation of the conservation equations

The definitions in § 4 of the quantities ϱ, s, and τ are purely formal and provide little physical insight into why these quantities are to be identified with the density, etc., of a fluid as we normally think of such things. The method used has the virtue of requiring almost no prior physical insight into fluids, but, on the other hand, it provides one with no subsequent insight either. In this section, a little physical understanding is assumed, and the conservation equations are derived again on that basis. By doing this, we are able not only to impart some significance to the quantities defined earlier, but also to interpret the term appearing on the right of (4.16) that so far has been ignored.

We consider now a continuous medium which we call a fluid and in which a density function $\varrho\,(t, q)$ is defined. This means that the mass of fluid in any domain V is simply the integral

$$\int_V \varrho\,(t, q)\,dq. \tag{7.1}$$

To any time t and any point q of the fluid, there is associated a velocity, which we denote by $s\,(t, q)$. A moving point $q = q(t)$ is said to *move with the fluid* if the function $q(t)$ is a solution of the differential equation

$$\frac{dq}{dt} = s\,(t, q). \tag{7.2}$$

If s is smooth, the initial point of a curve that moves with the fluid

$$q(0) = q^0 \tag{7.3}$$

can be transformed into the point $q\,(t, q^0)$ by solving the initial value problem (7.2-3). We say that a family of domains $V(t)$ *moves with the fluid* if each of its points is derived from a point $q^0 \in V(0)$ by the transformation

$$q = q\,(t, q^0).$$

If the domains $V(t)$ move with the fluid, then $V(t)$ always consists of the same "particles" of fluid. Since the mass of fluid in $V(t)$ is

$$\int_{V(t)} \varrho\,(t,\,q)\,dq, \tag{7.4}$$

mass is conserved if and only if the integral (7.4) is constant for any domain $V(t)$ moving with the fluid. Now, the transport theorem (2.10) shows that if $V(t)$ moves with the fluid, then

$$\frac{d}{dt}\int_{V(t)} \varrho\,dq = \int_{V(t)} (\varrho_t + \nabla\cdot(\varrho s))\,dq.$$

Mass is conserved if and only if this integral is zero. Since $V(t)$ is arbitrary, this means that mass is conserved if and only if

$$\varrho_t + \nabla\cdot(\varrho s) = 0, \tag{7.5}$$

and this is the conservation equation (4.7). It is the fact that (4.7) and (7.5) are identical that led us to define ϱ and s by (4.1) and (4.2) and to identify these quantities with the density and the velocity of a fluid.

To derive (4.16) from fluid mechanical considerations, a brief digression into the nature of stress is required, mainly because this idea is not as accessible to the intuition as are the ideas of density and velocity. In any fluid in motion, every portion of the fluid exerts forces on the neighboring portions. (If this were not the case, a pail of water that is initially stirred would never come to rest.) It is assumed that there exists a matrix function τ of t and q with the following property. Let D_r be a plane disk of radius r. Let n be the normal to D_r at its center. It is assumed that the force F_r exerted by the fluid across D_r has the form

$$F_r = \int_{D_r} \tau\cdot n\,dA.$$

The matrix τ is called the *stress tensor* associated with the fluid. The quantity $\tau\cdot n$ is called the *stress* across D_r. If V is a smooth domain in the fluid, then, approximating ∂V by its tangent plane at each point and summing over all the points on ∂V, one obtains the following formula for the internal force exerted on V by the rest of the fluid:

$$F_I = -\int_{\partial V} \tau\cdot n\,dA. \tag{7.6}$$

The *momentum* of the fluid in a domain V is defined to be $\int_V \varrho s\,dq$ — roughly, its mass times its velocity. Newton's second law says that the rate of change of momentum of the fluid in V must equal the sum of the forces on V. Now, the forces on a fluid are of two types: the internal forces (7.6), and any external forces (again, like gravity) acting on the fluid. The internal force (7.6) is written in terms of the stress, $\tau \cdot n$, which clearly is a pressure — that is, it has the dimensions of force per unit area. The external forces, on the other hand, are supposed to act per unit volume. This means that there exists a function $Q(t, q)$ such that the external forces on the fluid in V have the form

$$F_E = \int_V \varrho Q \, dq. \tag{7.7}$$

(7.6) and (7.7) show that Newton's law of conservation of momentum reads

$$\frac{d}{dt} \int_V \varrho s \, dq = - \int_{\partial V} \tau \cdot n \, dA + \int_V \varrho Q \, dq$$

$$= - \int_V \nabla \cdot \tau \, dq + \int_V \varrho Q \, dq, \tag{7.8}$$

by the divergence theorem. Letting $V(t)$ be a family of domains moving with the fluid, we can evaluate the left side of (7.8) by means of the transport theorem. Let s^i, $(\nabla \cdot \tau)^i$, and Q^i be the ith component of the vectors s, $(\nabla \cdot \tau)$, and Q, respectively. Then, we find

$$\int_{V(t)} [(\varrho s^i)_t + \nabla \cdot (\varrho s^i s) + (\nabla \cdot \tau)^i] \, dq = \int_{V(t)} \varrho Q^i \, dq.$$

Since $V(t)$ is arbitrary, we must have

$$(\varrho s^i)_t + \nabla \cdot (\varrho s^i s) + (\nabla \cdot \tau)^i = \varrho Q^i.$$

Now, using (7.5) to eliminate the term ϱ_t, we find, finally, that conservation of momentum is equivalent to

$$s_t + (s \cdot \nabla) s + \frac{1}{\varrho} \nabla \cdot \tau = Q. \tag{7.9}$$

The left side of (7.9) is the same as the left side of (4.16). This tells us how the term on the right of (4.16) must be interpreted. We must have

$$\int_{R^{6N-3}} Q_1 F \, d\zeta = \varrho Q,$$

Q being the external force per unit volume.

The term

$$\int_{R^{6N-3}} Q_1 F \, d\zeta$$

depends, through Q_1, on both the internal forces between the particles and the external forces on all of them. Presumably, the internal forces somehow cancel in the integral to leave just the external force, but I don't think this has ever really been explored.

As in § 4, if the fluid is incompressible, ϱ is constant, so that (7.5) becomes simply

$$\nabla \cdot s = 0. \tag{7.10}$$

CHAPTER 2

Potential flow

1 Ideal fluids

As WE POINTED out in § 1.5, there is no known system of particles for which
the Stokes hypothesis can be proved valid. But the experimentalists assure
us that real fluids satisfy the hypothesis. Well, if one cannot find theorems in
one place, perhaps he can find them in another. Since it is a mathematician's
business to prove theorems, the impasse can be resolved by simply assuming
the Stokes hypothesis. That is what we do here—with a vengeance. If one is
going to assume that σ is a function of D, the function assumed may as well
be simple, and the simplest function one can think of is the zero function. An
incompressible fluid for which the reduced stress tensor is identically zero is
called an *ideal fluid*. For ideal fluids,

$$\nabla \cdot \tau = \nabla p$$

(see (1.4.18)), so that the conservation equations (1.7.9) and (1.7.10) become

$$s_t + (s \cdot \nabla) s = -\frac{1}{\varrho} \nabla p + Q, \tag{1.1}$$

$$\nabla \cdot s = 0. \tag{1.2}$$

We shall see that these equations have many solutions. Therefore, other
conditions must be imposed on the solution if it is to be possible to select
a specific one. One appropriate condition is a consequence of the next
result.

THEOREM 1.1 *Consider a system of particles reflecting specularly off the walls
of a domain V. In such a system, the normal component of s on the boundary
of V is zero. In symbols,*

$$n \cdot s = 0 \quad for \quad q \in \partial V, \tag{1.3}$$

where n is the unit normal to ∂V.

If s is interpreted as the velocity of a flow, the normal component of s
on V—that is, $n \cdot s$—is just the velocity of the flow normal to the boundary.
To say that $n \cdot s = 0$, then, is to say neither more nor less than that no
fluid escapes from V.

To prove the theorem, we turn to the specular reflection condition (1.6.2). The definition of s is

$$\varrho s\,(t, q) = \int_{R^{6N-3}} ms_1 F\,(t, q, q_*, s_1, s_*)\,dq_*\,ds_*\,ds_1$$

$$= \int_{R^{6N-3}} ms_1 F\,(t, q, q_*, s_1 - 2\,(n \cdot s_1)\,n, s_*)\,dq_*\,ds_*\,ds_1$$

if $q \in \partial V$, by (1.6.2). Write $s_1 - 2\,(n \cdot s_1)\,n = \sigma_1$ in this integral. The result is

$$\varrho s\,(t, q) = \int_{R^{6N-3}} m\,[\sigma_1 - 2\,(n \cdot \sigma_1)\,n]\,F\,(t, q, q_*, \sigma_1, s_*)\,dq_*\,ds_*\,d\sigma_1$$

$$= \varrho s\,(t, q) - 2n \cdot [\varrho s\,(t, q)]\,n.$$

It follows that

$$n \cdot s\,(t, q) = 0 \quad \text{if} \quad q \in \partial V,$$

as desired.

Since (1.3) both follows from specular reflection and is such a reasonable condition physically, solutions of (1.2–2) satisfying (1.3) will be sought. It will be seen later on that (1.3) determines solutions of (1.1–2) uniquely.

2 The good fairy strikes!

In many cases of interest, the external forces, represented by Q in (1.1), can be derived from a potential. This means that there is a scalar function Ω such that

$$Q = \nabla\Omega. \tag{2.1}$$

This is the case in the absence of external forces when, of course, we can take $\Omega = 0$. It is also the case when the external force is due to gravity, when $\Omega = -gz$, z being the vertical coordinate and g the acceleration due to gravity. In most of what follows, (2.1) is assumed. Then, (1.1–2) become

$$s_t + (s \cdot \nabla)\,s = -\frac{1}{\varrho}\,\nabla p + \nabla\Omega, \tag{2.2}$$

$$\nabla \cdot s = 0. \tag{2.3}$$

The simplest vector function that one might think of is the gradient of a scalar. After all, such a vector function depends not on three, but only on one, independent scalar function. The simplest possible solution of (2.2–3) then might have the form

$$s = \nabla\varphi. \tag{2.4}$$

As we shall see, the substitution (2.4)—which represents no more than a quest for solutions of an exceptionally simple form—succeeds astonishingly well in producing solutions and insight into solutions of (2.2–3).

A straightforward calculation shows that if s is given by (2.4), then

$$(s \cdot \nabla) s = \tfrac{1}{2} \nabla (|\nabla \varphi|^2).$$

Therefore, (2.2) becomes

$$\nabla \left(\varphi_t + \frac{1}{2} |\nabla \varphi|^2 + \frac{1}{\varrho} p \right) = \nabla \Omega.$$

This can be integrated to give

$$\varphi_t + \frac{1}{2} |\nabla \varphi|^2 + \frac{1}{\varrho} p = \Omega + c(t),$$

where $c(t)$ is a function of t *alone*. Since, as is clear from (2.2), the addition of a function of t to p changes nothing in the flow, we may absorb $c(t)$ in p and write

$$\varphi_t + \frac{1}{2} |\nabla \varphi|^2 + \frac{1}{\varrho} p = \Omega. \tag{2.5}$$

Equations (2.5) is called *Bernoulli's equation*, and, if the function φ is given, it can be used to determine the pressure.

If s is a gradient, Bernoulli's equation is equivalent to (2.2). As for (2.3), when s is given by (2.4), (2.3) will be satisfied if

$$\nabla^2 \varphi = \varphi_{xx} + \varphi_{yy} + \varphi_{zz} = 0. \tag{2.6}$$

Any solution of this equation is called a *harmonic function*. What we have proved is

THEOREM 2.1   *Let φ be any harmonic function depending smoothly on a parameter t. Then, the functions s and p defined by (2.4) and (2.5), respectively, satisfy (2.2–3).*

The function φ of the theorem is called the *velocity potential* of the flow. Flows of the type described are called *potential flows*. That potential flows exist represents a stroke of extraordinary good luck. Not only do these flows depend upon the single function φ instead of on four functions, but φ itself satisfies the *linear* differential equation (2.6). It is hard to overestimate the simplicity that has been injected into our studies because of this fact.

3 Lagrange's theorem

We shall use theorem 2.1 shortly to derive certain simple facts about potential flows. But first, it is appropriate to ask when potential flows arise. The result that gives the answer is sometimes called *Lagrange's theorem*.

THEOREM 3.1   *Let $s(t, q)$ be a solution of (2.2–3) satisfying*

$$s(t_0, q) = \nabla\varphi_0(q) \tag{3.1}$$

for some fixed t_0. Then, $s(t, q)$ is derivable from a potential for all values of t — that is, there exists a function $\varphi(t, q)$ such that

$$s(t, q) = \nabla\varphi(t, q).$$

This says that any flow that begins as potential flow remains so always. In particular, we have the following important result for flows that begin from rest.

COROLLARY 3.2   *Let $s(t, q)$ be any solution of (2.2–3) satisfying*

$$s(t_0, q) = 0.$$

Then, there exists a velocity potential for the flow.

The corollary follows immediately from the theorem by setting $\varphi_0 = 0$.

In the proof of the theorem, we use a quantity associated with any simple closed curve in a fluid. Let s be a solution of (2.2–3) in a domain V, and let γ be a simple closed curve in V. Then, the line integral

$$\Gamma = \int_\gamma s \cdot dq = \int_\gamma (u\,dx + v\,dy + w\,dz)$$

is called the *circulation* of the flow around γ.

Let us say that a simple closed curve γ is *immersed* in V if there exists a surface S, whose boundary is γ, and which consists entirely of points of V. If V is simply connected, every simple closed curve in V is immersed. But in flows past a circular cylinder, say, no curve that surrounds the cylinder is immersed.

The reason these concepts have been introduced is to allow us to state

LEMMA 3.3   *Let $s(t, q)$ be a solution of (2.2–3) in a domain V. For any fixed t, $s(t, q)$ is derivable from a potential if and only if the circulation is zero around any simple closed curve immersed in V.*

To prove lemma 3.3, let γ be any simple closed curve immersed in V, and let S be a surface bounded by γ lying entirely in V. If n denotes the normal to S, Stokes' theorem shows that

$$\Gamma = \int_\gamma s \cdot dq = \int_S n \cdot (\nabla \times s)\, dA, \qquad (3.2)$$

dA being the area element on S.

Now, if s has the form $\nabla \varphi$ for some fixed t, a trivial computation shows that $\nabla \times s = 0$, so that the circulation is zero.

Conversely, if the circulation is zero about every curve immersed in the fluid, (3.2) shows that $\nabla \times s = 0$ at every point of V (since γ is an arbitrary immersed curve in V), and this implies that s is a gradient. This proves lemma 3.3.

Next, we recall a definition made in chapter one. A curve $q(t)$ is said to *follow the fluid* if it is solution of the differential equation

$$\frac{dq}{dt} = s\,(t, q). \qquad (3.3)$$

Roughly, a curve follows the fluid if it is traced out by a "particle of fluid" as time progresses. A family of simple closed curves $\gamma(t)$ are said *follow the fluid* if the curve traced out by each of its points does so.

We can now state

LEMMA 3.4   (Kelvin). *Let $\gamma(t)$ be a simple closed contour that follows the fluid. Then the circulation around $\gamma(t)$ is constant.*

Proof   Since $\gamma(t)$ follows the fluid, it can be parametrized in the form

$$q = q\,(\tau, t) \qquad (3.4)$$

where, for each t, $0 \leqslant \tau \leqslant 1$, the curve (3.4) is simple and closed. Moreover, for fixed τ, the curve $q = q\,(\tau, t)$ follows the fluid.

By definition, the circulation around $\gamma(t)$ is given by

$$\Gamma(t) = \int_0^1 s \cdot q_\tau\, d\tau.$$

Therefore, if we denote time derivatives by a dot, we have

$$\dot{\Gamma}(t) = \int_0^1 (\dot{s} \cdot q_\tau + s \cdot q_{\tau t})\, d\tau = \int_0^1 \{[s_t + (s \cdot \nabla)\, s] \cdot q_\tau + s \cdot s_\tau\}\, d\tau$$

since for fixed τ, $q\,(\tau, t)$ moves with the fluid.

By (2.2), we have

$$\dot{\Gamma}(t) = \int_0^1 \left[\nabla \left(\Omega - \frac{1}{\varrho}\, p \right) \cdot q_\tau + s \cdot s_\tau \right] d\tau$$

$$= \int_0^1 \frac{\partial}{\partial \tau} \left[\Omega - \frac{1}{\varrho}\, p + \frac{1}{2}\, |s|^2 \right] d\tau = 0,$$

since $\Omega - (1/\varrho)p + \frac{1}{2}|s|^2$ takes on the same value at both ends of $\gamma(t)$. This completes the proof of Kelvin's lemma.

It is now an easy matter to prove Lagrange's theorem. If $s(t_0, q) = \nabla\varphi_0(q)$, then the necessity portion of lemma 3.3 shows that $\Gamma(t_0) = 0$. Then, lemma 3.4 shows that $\Gamma(t) = 0$ for all t. The sufficiency portion of lemma 3.3 then states that $s(t, q)$ is derivable from a potential for all t. This completes the proof.

It is interesting to notice that the proof of Lagrange's theorem shows that the hypothesis (3.1) need not hold throughout the entire fluid. In a portion of the fluid, the velocity vector s may initially be a gradient; in the rest of the fluid it need not be. The first portion will then carry the property of being a gradient along with it and thus need never be confused with the rest. *No portion of the fluid that is not initially a gradient ever acquires that property, and no portion that has it initially ever loses it.*

Exercises

3.1   Show in detail why, as in the proof of lemma 3.3, $\nabla \times s$ is zero when Γ is zero.
3.2   Show that $\nabla \times s = 0$ implies $s = \nabla\varphi$.

4   Some examples of potential flow

According to theorem 2.1, any solution of Laplace's equation (2.6) gives rise to a solution of equations (2.2–3). Fortunately, a good many solutions of (2.6) are known. We examine some of these and discuss briefly some properties of the corresponding flows.

(1)   One solution of Laplace's equation is the simple

$$\varphi^{(1)} = Ux, \tag{4.1}$$

where U is a constant. Since $\varphi_y^{(1)}$ and $\varphi_z^{(1)}$ are identically zero, while $\varphi_x^{(1)} \equiv U$, the associated velocity vector has the components

$$u = U, \quad v = w = 0,$$

and the flow is a uniform one in the direction of the x-axis.

Since the components of the velocity vector in the y and the z directions are zero, any plane $y = $ constant or $z = $ constant may be regarded as a rigid wall through which no fluid flows. Thus, (4.1) represents the steady potential flow down any channel with a rectangular cross-section lying parallel to the x-axis. This flow is called *uniform flow*.

(2)   Introduce spherical polar coordinates by means of the equations

$$\left.\begin{aligned} x &= r \cos \omega, \\ y &= r \sin \omega \cos \theta, \\ z &= r \sin \omega \sin \theta. \end{aligned}\right\} \tag{4.2}$$

A direct computation shows that a solution[1] of (2.6) is

$$\varphi^{(2)} = \frac{1}{r} \cos \omega. \tag{4.3}$$

The solution (4.3) is not too interesting in itself. However (2.6) is a *linear* equation, and (4.1) and (4.3) are solutions of it. Therfore, any linear combination of (4.1) and (4.3) is a solution. Consider, then the solution

$$\varphi^{(3)} = U \left(x + \frac{\alpha}{r} \cos \omega \right) \tag{4.4}$$

$$= U \left(r + \frac{\alpha}{r} \right) \cos \omega \tag{4.5}$$

where α is any constant.

We have seen that the velocity vector associated with a velocity potential φ is

$$s = \nabla \varphi.$$

If we have a rigid wall with normal vector n, we require

$$n \cdot s = n \cdot \nabla \varphi = 0$$

on the wall. However, $n \cdot \nabla \varphi$ is the directional derivative in the direction of n. Thus, for potential flow, the requirement that there be no flow through a rigid wall is equivalent to the condition

$$\frac{\partial \varphi}{\partial n} = 0, \tag{4.6}$$

the symbol $\partial / \partial n$ denoting differentiation in the direction normal to the wall.

From (4.5), we have

$$\frac{\partial \varphi^{(3)}}{\partial r} = U \left(1 - \frac{\alpha}{r^2} \right) \cos \omega.$$

This will be zero when $r = r_0$ if $\alpha = r_0^2$. Thus, the function

$$\varphi^{(3)} = U \left(r + \frac{r_0^2}{r} \right) \cos \omega \tag{4.7}$$

---

[1] Not really obtained by magic, but by separating variables in the coordinate system (4.2).

is the velocity potential of a flow past a rigid sphere centered at the origin and having radius r_0.

Moreover, (4.7) shor that

$$\varphi^{(3)} = (Ux, 0, 0) + 0(1) \quad \text{as} \quad r \to \infty.$$

Thus, at infinity, the flow described by (4.7) tends to a uniform flow in the direction of the x-axis.

A sketch of the curves that move with the fluid is given in figure 2. These curves are the same in all planes $\theta = $ constant and, in such plane, the equations of the curves (in polar coordinates) are

$$\left(r - \frac{1}{r} \right) \sin \omega = c,$$

where c is a constant. The curves appearing in Figure 2, were sketched using this equation. The method by which this equation is derived is discussed in exercise 4.1 below.

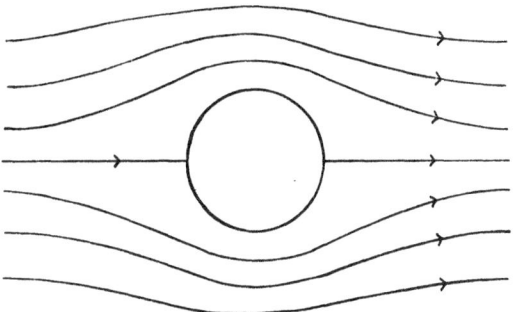

Figure 2

(3) An example of a flow that *cannot* be derived from a potential is one with uniform angular velocity about the z-axis. For this flow

$$u = -\omega y,$$

$$v = \omega x, \tag{1.8}$$

$$w = 0.$$

where ω is a constant. s cannot be derived from a potential since $u_y \neq v_x$. However, if we assume that the external forces are derived from a potential Ω, then the velocity vector defined by (4.8) satisfies (2.2–3) if the pressure is defined by

$$\frac{1}{\varrho}p = \Omega + \frac{1}{2}\omega^2 (x^2 + y^2). \tag{4.9}$$

Suppose the only external force is that due to gravity. Choosing the z-axis to point *up*, we obtain $\Omega = -gz$. We then have a mass of fluid, rotating under gravity with constant angular velocity ω about a vertical axis. One can think of all this as a rotating whirlpool of water. At the free surface of the water, the pressure must equal atmospheric pressure, for obvious reasons. Therefore, (4.9) gives for the shape of the free surface

$$gz = \tfrac{1}{2}\omega^2 (x^2 + y^2) + \text{constant}, \tag{4.10}$$

which is the equation of a paraboloid of revolution.

(4)   Of course, (4.10) does not represent a whirlpool very realistically since, obviously, the height of the free surface increases without bound as x and y go to infinity.

The difficulty may be that we are requiring too much by insisting that the angular velocity ω be constant. However, one might expect a realistic hypothesis to be that the angular velocity depends only on the distance from the axis of rotation.

Introduce cylindrical coordinates

$$\left. \begin{array}{l} x = r \cos \theta \\ y = r \sin \theta, \end{array} \right\} \tag{4.11}$$

and suppose that

$$\begin{aligned} u &= -\omega y \\ v &= \omega x \\ w &= 0, \end{aligned} \tag{4.12}$$

while ω is a function of r alone. Any flow in which $w = 0$ is derivable from a potential if and only if $v_x - u_y = 0$. However, in the present case, we have

$$v_x - u_y = 2\omega + r \frac{d\omega}{dr},$$

and this can only be zero if

$$\omega = \frac{\mu}{r^2}, \tag{4.13}$$

where μ is a constant.

There is a velocity potential associated with the flow described by (4.11) and (4.13). A direct computation shows that the function in question is

$$\varphi^{(4)} = -\mu\theta + \text{constant}. \tag{4.14}$$

The associated pressure can be found from Bernoulli's equation. If $\Omega = -gz$, we have

$$\frac{1}{\varrho} p = -gz - \frac{1}{2} \frac{\mu^2}{r^2} + \text{constant}.$$

The free surface, where the pressure is constant, has the form

$$gz = \text{constant} - \frac{\mu^2}{2r^2}. \tag{4.15}$$

(5)   To illustrate a procedure and a type of argument used often in fluid mechanics, we carry our discussion of "whirlpools" slightly further.

Neither of the flows (4.8) nor (4.9) is completely satisfactory as a description of the situation. The free surface associated with the flow (4.8) is a paraboloid, and so blows up as $r \to \infty$. The free surface associated with (4.14), on the other hand, is given by (4.15), and this blows up when $r \to 0$.

Rankine suggested putting the two flows together in a continuous way; the result he called a *combined vortex*. We describe this vortex formally first and attempt to justify the formulas obtained afterwards.

One assumes that for $0 < r < a$, the flow is given by

$$u = -\omega y$$
$$v = \omega x \tag{4.16}$$
$$w = 0,$$

with ω constant, while for $r > a$, it is given by

$$u = - \frac{\mu y}{r^2}$$
$$v = \frac{\mu x}{r^2} \tag{4.17}$$
$$w = 0.$$

Take

$$\mu = a^2.$$

Then, the velocity is continuous throughout the flow. The free surface is described by equations of the form

$$gz = \begin{cases} \dfrac{\omega^2}{2}(r^2 - a^2) + c, & 0 < r < a, \\[2ex] \dfrac{\omega^2}{2}\left(a^2 - \dfrac{a^4}{r^2}\right) + c, & r > a. \end{cases}$$

A graph of the free surface is shown in Figure 3. The minimum depth below the depth at infinity is $(\omega^2 a^2)/g$.

An heuristic justification of Rankine's combined vortex is this. It is often found experimentally that flows are very nearly potential flows when the velocities are small. Away from the axis of rotation ($r = 0$) the velocities (4.17) are small. Therefore, it might be expected that the approximation of potential flow is valid there. Near $r = 0$, on the other hand, the velocities associated with (4.17) are large; therefore, the layers of fluid tend to stick together and it rotates like a solid body, with constant angular velocity. Consequently, near $r = 0$, the flow should resemble that described by (4.16).

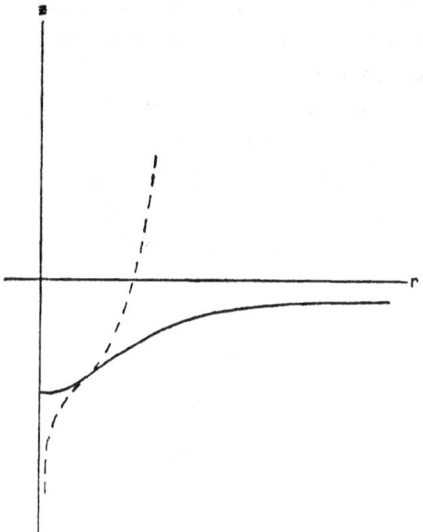

Figure 3

That is the argument, for what it is worth. We shall meet a more sophisticated version of it later when we speak of boundary layers.

Exercise

4.1 Show that in spherical polar coordinates (4.2) the differential equations for the curves that move with the fluid in example (2) are

$$\frac{dr}{dt} = \left(1 - \frac{1}{r^2}\right)\cos\omega$$

$$r\frac{d\omega}{dt} = -\left(1 + \frac{1}{r^2}\right)\sin\omega$$

$$\frac{d\theta}{dt} = 0.$$

Conclude from this that the curves that move with the fluid are the same in all planes $\theta = $ constant and that in such a plane they have the equations

$$\left(r - \frac{1}{r}\right)\sin\omega = \text{constant}.$$

CHAPTER 3

Some properties of potential flows

1 Introduction

SINCE THERE is much to learn from inspection of examples, we shall have more of them in the next chapter. But first, we derive certain useful properties of potential flows.

A basic fact in all that follows is *Green's theorem*, which says that if V is any domain with a nice enough boundary[1], and if f and g are any twice differentiable functions in V with one continuous derivative in the closure of V, then

$$\int_V (f\nabla^2 g + \nabla f \cdot \nabla g)\, dq = \int_{\partial V} f \frac{\partial g}{\partial n}\, dA. \tag{1.1}$$

Here ∂V denotes the boundary of V, $\partial/\partial n$ denotes differentiation in the direction of the exterior normal to ∂V, and dA denotes the area element on ∂V.

In the rest of this chapter, φ denotes any solution of

$$\nabla^2 \varphi = 0 \tag{1.2}$$

in a domain V with a sufficiently smooth boundary. By a solution, we mean, in this chapter at least, a function with two continuous derivatives in V, whose first derivatives are continuous in the closure of V, and which satisfies (1.2).

Theorem 2.2.1 shows that if the external forces are derivable from a potential φ, then there is a velocity field associated with every solution of (1.2), given by

$$s = \nabla\varphi. \tag{1.3}$$

The pressure in the fluid moving with the velocity (1.3) is then given by Bernoulli's equation

$$\frac{1}{\varrho} p = \Omega - \varphi_t - \frac{1}{2}|\nabla\varphi|^2. \tag{1.4}$$

In all that follows, we denote the ball of radius r about a point $q_0 \in R^3$ by $B_r(q_0)$,

$$B_r(q_0) = \{q \in R^3 : |q - q_0| < r\}. \tag{1.5}$$

---

[1] All the domains we apply the theorem to have such a boundary.

The sphere which is the boundary of $B_r(q_0)$ is denoted by $S_r(q_0)$,

$$S_r(q_0) = \{q \in R^3: |q - q_0| = r\}. \tag{1.6}$$

2   Gauss's theorem

Let φ be any solution of (1.2). Set $f = 1$ and $g = \varphi$ in (1.1). The conclusion is that

$$\int_{\partial V} \frac{\partial \varphi}{\partial n} \, dA = 0. \tag{2.1}$$

Let V be the ball $B_r(q)$ of radius r centered at the point q. Introduce spherical polar coordinates with the origin at q:

$$\xi = x + r \cos \omega$$

$$\eta = y + r \sin \omega \cos \theta$$

$$\zeta = z + r \sin \omega \sin \theta.$$

Then, $dA = r^2 \sin \omega \, d\theta \, d\omega$, $\partial/\partial n = \partial/\partial r$, and (2.1) becomes

$$0 = \int_0^\pi \sin \omega \, d\omega \int_0^{2\pi} \frac{\partial}{\partial r} \varphi \times$$

$$\times (x + r \cos \omega, y + r \sin \omega \cos \theta, z + r \sin \omega \sin \theta) \, d\theta$$

$$= \frac{\partial}{\partial r} \int_0^\pi \sin \omega \, d\omega \int_0^{2\pi} \varphi \times$$

$$\times (x + r \cos \omega, y + r \sin \omega \cos \theta, z + r \sin \omega \sin \theta) \, d\theta.$$

This equation says that the integral occurring in it is a constant. Letting $r \to 0$, the integral becomes

$$\varphi(q) \int_0^\pi \int_0^{2\pi} \sin \omega \, d\theta \, d\omega = 4\pi\varphi(q).$$

Thus, we have

$$\varphi(q) = \frac{1}{4\pi} \int_0^\pi \int_0^{2\pi} \varphi (x + r \cos \omega, y + r \sin \omega \cos \theta,$$

$$z + r \sin \omega \sin \theta) \sin \omega \, d\omega d\theta.$$

The right hand side of this formula is the mean value of φ over the sphere. Thus, we have

THEOREM 2.1   (Gauss). *Let φ be any solution of* (1.2) *in a ball $B_r(q)$ of radius r and center q. Then, φ at q is equal to its mean value over the boundary of $B_r(q)$. Briefly*

$$\varphi(q) = \frac{1}{4\pi r^2} \int_{S_r(q)} \varphi \, dA.$$

Exercise

2.1   Let φ be a function defined on a domain V in R^2 instead of R^3. Show that the analog of Gauss's theorem holds here, too. That is, prove that if φ is any harmonic function of two variables, then its value at the center of any disk is equal to its mean value on the boundary of the disk.

3   The maximum principle

With the aid of Gauss's theorem, we can now easily prove the following result, usually called the *maximum principle.*

THEOREM 3.1   *Let $\varphi(q)$ be a solution of* (1.2) *in a domain V. Then φ cannot take on either a maximum or a minimum at an interior point of V unless it is identically constant.*

For, suppose φ had a maximum (say) at a point q_0 in V. Then,

$$\varphi(q_0) \geqslant \varphi(q)$$

for every point q in a ball $B_r(q_0)$ centered at q_0. (2.2) shows that this is impossible unless $\varphi(q) = \varphi(q_0)$ for every q in the ball. Therefore, $\varphi(q)$ is constant on the ball. It follows that $\varphi(q)$ is constant on every ball contained in V whose center lies in $B_r(q_0)$. For if not, φ would have a maximum on the boundary of the new ball, and the argument could be repeated. Since V is connected (see § 1.1), it can be covered entirely by overlapping balls in each of which the argument can again be repeated. Thus, $\varphi(q) = \varphi(q_0)$ in all of V, and the result follows.

There is also a maximum principle for the speed $|\nabla \varphi|$, but this one differs from the maximum principle for φ in that there is no related minimum principle.

THEOREM 3.2   *Let $\varphi(q)$ be a solution of* (1.2) *in a domain V. Then, $|\nabla \varphi|$ cannot take on a maximum at an interior point of V unless it is identically constant. This result is false if "maximum" is replaced by "minimum".*

The proof is as follows. Let q_0 be a point of V at which the putative maximum occurs. Rotate the coordinate system so that the x-axis is in the direction of the velocity vector at q_0. Then,

$$s(q_0) = (\nabla\varphi)(q_0) = (\varphi_x(q_0), 0, 0).$$

Since φ satisfies (1.2), so does φ_x. Therefore, the maximum principle applies to φ_x to show that it has neither a maximum nor a minimum in V. Unless φ_x is identically constant, then, $\varphi_x{}^2$ has no maximum in V. Therefore, in every neighborhood of the point q_0, there exist points q at which

$$\varphi_x{}^2(q) > \varphi_x{}^2(q_0).$$

Trivially, then

$$|\nabla\varphi(q)|^2 \geqslant \varphi_x{}^2(q) > \varphi_x{}^2(q_0) = |\nabla\varphi(q_0)|^2,$$

and $|\nabla\varphi|^2$ does not have a maximum at q_0 unless φ_x is identically constant. If φ_x is constant, though, $\varphi = ax + \varphi'$, where a is constant and φ' is a function of y and z alone, satisfying

$$\varphi'_{yy} + \varphi'_{zz} = 0.$$

Still assuming $|\nabla\varphi|^2$ has a maximum at q_0, we see that $|\nabla\varphi'|^2$ must have a maximum there too, since $|\nabla\varphi|^2 = a^2 + |\nabla\varphi'|^2$. But we have so adjusted the coordinate system that $|\nabla\varphi'| = 0$ at q_0. Therefore, $|\nabla\varphi'|$ must be identically zero, and $|\nabla\varphi|^2 \equiv a^2$, a constant. This proves the first part of theorem 2.

To show that $|\nabla\varphi|^2$ *can* have a minimum, introduce cylindrical coordinates

$$x = r\cos\theta$$

$$y = r\sin\theta,$$

and set

$$\varphi = \frac{A}{r}\cos\theta - B\theta, \tag{3.1}$$

where A and B are constants. We shall discuss the function (3.1) in some detail later. For now, we merely note that it satisfies (1.2) in any domain that does not include $r = 0$, while $|\nabla\varphi|^2 = 0$ at the point $r = A/B$, $\theta = -\pi/2$. Since $|\nabla\varphi|^2$ is clearly never negative, this completes the proof of theorem 2.

4   The minimum principle for the pressure

One expects a certain one-sidedness connected with the pressure. Although it is easy to believe that pressure increases without bound as one goes deeper and deeper in an ocean of infinite depth, it is hard to see how pressure can become less than that in a hard vacuum.

If there are no external forces and φ is independent of t, the one-sided aspect of the pressure is an immediate consequence of theorem 3.2. For, according to Bernoulli's equation, the pressure is given by[1]

$$\frac{1}{\varrho} p = - \frac{1}{2} |\nabla\varphi|^2. \tag{4.1}$$

Since $|\nabla\varphi|^2$ cannot have an interior maximum, this shows that p cannot have an interior minimum.

But the result is more general than this. We have, in fact,

THEOREM 4.1 *Let p be the pressure associated with a potential flow in a domain V. Suppose that the potential Ω of the external forces satisfies Laplace's equation[2]* :

$$\nabla^2\Omega = 0.$$

Then, p cannot take on a minimum at an interior point of V unless the flow is uniform[3]. This result is false if "minimum" is replaced by "maximum".

Let φ be the velocity potential associated with the flow. Then, since φ satisfies (1.2), so does φ_x. Applying (1.1) with $f = g = \varphi_x$, we find

$$\int_{\partial W} \varphi_x \frac{\partial\varphi_x}{\partial n} \, dA = \int_W |\nabla\varphi_x|^2 \, dq;$$

here, W is any subdomain of V. As in § 2, let W be $B_r(q)$, the ball of radius r about any point q of V. We have

$$dA = r^2 \sin\omega \, d\omega \, d\theta = r^2 \, d\alpha,$$

say. Therefore,

$$\frac{r^2}{2} \frac{d}{dr} \int_{S_r(q)} \varphi_x^2 \, d\alpha = \frac{1}{2} \int_{S_r(q)} \frac{\partial}{\partial r} (\varphi_x^2) \, dA = \int_{S_r(q)} \varphi_x \varphi_{xr} \, dA$$

$$= \int_{B_r(q)} |\nabla\varphi_x|^2 \, dA > 0,$$

unless φ_x is constant in $B_r(x)$. A similar inequality holds with φ_x replaced by φ_y and φ_z. Therefore, we see that unless $\nabla\varphi$ is constant—that is, unless the flow is uniform,

$$\frac{1}{4\pi} \int_{S_r(q)} |\nabla\varphi|^2 \, d\alpha = \frac{1}{4\pi r^2} \int_{S_r(q)} |\nabla\varphi|^2 \, dA$$

is a strictly increasing function of r.

---

[1] (4.1) does not imply the pressure is negative since a constant can be added to the right side without affecting equations (2.2–3).

[2] It should be noted that this is true when $\Omega = -gz$, the potential associated with external gravitational forces.

[3] That is, unless the velocity is everywhere constant.

Now Bernoulli's equation gives

$$\frac{1}{\varrho} p = \Omega - \varphi_t - \frac{1}{2} |\nabla\varphi|^2.$$

Since both Ω and φ_t satisfy Laplace's equation, the first two terms on the right have the property that their mean values over spheres are independent of the radii of the spheres (Gauss's theorem). By what we have just proved, then, unless the flow is uniform, the mean value of p over any sphere of radius r is a strictly *decreasing* function of r if the flow is not uniform. An easy consequence of this is that p cannot have an interior minimum.

That the theorem is false if "minimum" is replaced by "maximum" is shown by the example (3.1), where the pressure has a maximum at the point $r = A/B$, $\theta = -\pi/2$.

5   A variational principle for potential flows

If $s (t, q)$ is a velocity field associated with the flow of an incompressible fluid in a domain V, we define the *kinetic energy T* of the flow by the formula

$$T = \frac{\varrho}{2} \int_V |s|^2 \, dq. \tag{5.1}$$

We prove that potential flows are characterized by a minimum principle associated with the kinetic energy.

THEOREM 5.1   *Let φ be the velocity potential of a flow in a bounded domain V. Let $s (t, q)$ be the velocity field of any other flow in V satisfying the following two conditions:*

(i)   *the normal component of s is equal to the normal component of the potential flow associated with φ on the boundary of V, i.e.,*

$$n \cdot s = \frac{\partial\varphi}{\partial n} \quad on \quad \partial V;$$

(ii)   *s satifies the continuity equation*

$$\nabla \cdot s = 0 \quad in \ V.$$

Then,

$$\int_V |s|^2 \, dq \geqslant \int_V |\nabla\varphi|^2 \, dq. \tag{5.2}$$

Thus, potential flow has the least kinetic energy of any flow having the same normal velocity on the boundary and satisfying the condition of conservation of mass. It should be noted that the theorem characterizes potential flows among *all* flows satisfying these conditions; the flow associated with s need not even be ideal.

To prove the theorem, consider

$$\int_V |s - \nabla\varphi|^2 \, dq = \int_V |s|^2 \, dq + \int_V |\nabla\varphi|^2 \, dq - 2 \int_V s \cdot \nabla\varphi \, dq. \quad (5.3)$$

Now, using (ii), we find

$$\int_V s \cdot \nabla\varphi \, dq = \int_V \nabla \cdot (\varphi s) \, dq = \int_{\partial V} n \cdot \varphi s \, dA,$$

by the divergence theorem. But,

$$\int_{\partial V} n \cdot \varphi s \, dA = \int_{\partial V} \varphi \, (n \cdot s) \, dA = \int_{\partial V} \varphi \, \frac{\partial \varphi}{\partial n} \, dA = \int_V |\nabla\varphi|^2 \, dq$$

by (1.1) with $f = g = \varphi$.

Thus, according to (5.3),

$$\int_V |s - \nabla\varphi|^2 \, dq = \int_V |s|^2 \, dq - \int_V |\nabla\varphi|^2 \, dq. \quad (5.4)$$

Since the left side of this equation is clearly non-negative, (5.2) follows.

Notice that a slightly stronger result is also valid. (5.4) shows that the kinetic energy associated with the field s is *strictly greater* than that of the potential flow unless $s = \nabla\varphi$ everywhere in V.

6   Uniqueness of potential flows

We mentioned in § 2.1 that the condition (2.1.3) determines an ideal flow uniquely. The result is most important, for it says that for ideal fluids anyway, the specular reflection hypothesis is enough to pick out a flow with given initial conditions from all possible (ideal) flows satisfying these initial conditions.

The result is proved in this section for potential flows, and then in the next section for general ideal flows. Before going on, we recall that for potential flows, (2.1.3) is equivalent to the condition

$$\frac{\partial\varphi}{\partial n} = 0 \quad \text{for} \quad q \in \partial V. \quad (6.1)$$

If the domain occupied by the fluid is bounded, it is very easy to prove that the solution satisfying (6.1) is unique. In fact, it is even easy to prove that two solutions having the same normal derivative on ∂V (whether or not it is zero) are equal. For let $\varphi^{(1)}$ and $\varphi^{(2)}$ be two velocity potentials in a domain V, satisfying

$$\frac{\partial \varphi^{(1)}}{\partial n} = \frac{\partial \varphi^{(2)}}{\partial n} \quad \text{on} \quad \partial V.$$

Since (1.2) is linear, the difference

$$\varphi^{(1)} - \varphi^{(2)} \equiv \varphi$$

also satisfies (1.2) and, moreover, φ satisfies (6.1). Now, use (1.1) with $f = g = \varphi$. The result is

$$\int_V |\nabla \varphi|^2 \, dq = \int_V \varphi \, \frac{\partial \varphi}{\partial n} \, dA = 0.$$

Thus, $\nabla \varphi = 0$, and the velocity fields associated with the two potentials $\varphi^{(1)}$ and $\varphi^{(2)}$ are identical.

We need another hypothesis if V is unbounded. To state the result, we define what we mean by an exterior domain. A domain V is an *exterior domain* if it contains the exterior of a ball in its interior—that is, if there is an $r > 0$ such that

$$V \supset \{q \in R^3 \colon |q| > r\}.$$

THEOREM 6.1   *Let V be an exterior domain. Let $\varphi^{(1)}$ and $\varphi^{(2)}$ be two velocity potentials in V satisfying*

$$\frac{\partial \varphi^{(1)}}{\partial n} = \frac{\partial \varphi^{(2)}}{\partial n} \quad \text{on} \quad \partial V,$$

and

$$|\nabla \varphi^{(1)} - \nabla \varphi^{(2)}| \to 0 \quad \text{as} \quad |q| \to \infty.$$

Then,

$$\nabla \varphi^{(1)} \equiv \nabla \varphi^{(2)} \quad \text{in} \quad V.$$

This can also be stated as follows: there is only one potential flow over a body (the complement of V) having prescribed normal velocity on its boundary and prescribed velocity at infinity.

Let $\varphi = \varphi^{(1)} - \varphi^{(2)}$. Then, the hypotheses are that $\partial \varphi / \partial n = 0$ on ∂V and $|\nabla \varphi| \to 0$ at infinity. We have to prove that $\nabla \varphi \equiv 0$ in V. We prove first that φ goes to a constant at infinity.

LEMMA 6.2   *Let V be an exterior domain. Let φ be a velocity potential satisfying the two conditions*

$$\frac{\partial \varphi}{\partial n} = 0 \quad \text{on} \quad \partial V, \tag{6.2}$$

and

$$|\nabla\varphi| \to 0 \quad as \quad |q| \to \infty. \tag{6.3}$$

Then, $\varphi(q)$ approaches a constant as $|q| \to \infty$.

Let W be the complement of V. Since $|\nabla\varphi| \to 0$, while W is bounded, there is a ball $B_{r_0}(0)$ with r_0 so large that $W \subset B_{r_0}(0)$, and $|\nabla\varphi|$ is less than any prescribed number ε in the exterior of $B_{r_0}(0)$:

$$|\nabla\varphi| < \varepsilon \quad for \quad |q| > r_0. \tag{6.4}$$

Take $\delta > 0$, and let q be any point far enough away from the origin that the solid angle subtended at q by $B_{r_0}(0)$ is less than δ. Let $B_{r_1}(q)$ and $B_{r_2}(q)$ be two balls about q that are such that W lies outside $B_{r_1}(q)$ and inside $B_{r_2}(q)$. (See Figure 4.)

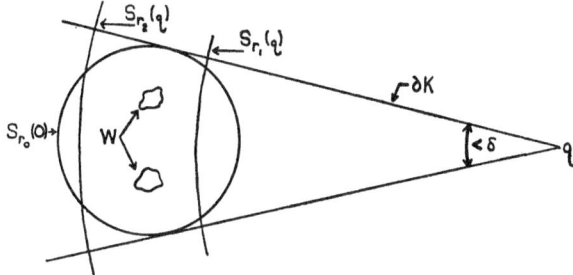

Figure 4

As we saw in § 2,

$$\varphi(q) = \frac{1}{4\pi r_1^2} \int_{S_{r_1}(q)} \varphi \, dA. \tag{6.5}$$

Notice that this formula is not correct if we replace r_1 by r_2, since φ is not harmonic in all of $B_{r_2}(q)$. However, (6.5) shows that

$$\varphi(q) = \frac{1}{4\pi r_2^2} \int_{S_{r_2}(q)} \varphi \, dA + \left[\frac{1}{4\pi r_1^2} \int_{S_{r_1}(q)} \varphi \, dA - \frac{1}{4\pi r_2^2} \int_{S_{r_2}(q)} \varphi \, dA \right]. \tag{6.6}$$

Let α be a unit vector, so that on a sphere of radius r, $dA = r^2 d\alpha$. The quantity in brackets in the above expression for φ is then

$$\frac{1}{4\pi} \int_{S_1(0)} [\varphi(q + r_1\alpha) - \varphi(q + r_2\alpha)] \, d\alpha.$$

Let K be the solid cone with vertex q such that ∂K is tangent to $B_{r_0}(0)$. Let D be the domain bounded by ∂K, $S_{r_2}(q)$, $S_{r_1}(q)$, and ∂V. As $q + r_1\alpha$ varies over $S_{r_i}(q) \cap K$ $(i = 1, 2)$, α varies over a small set Σ on the unit sphere of measure less than δ. Therefore,

$$\left| \int_{\Sigma} [\varphi (q + r_1\alpha) - \varphi (q + r_2\alpha)] \, d\alpha \right| \leqslant 2\delta \sup_{p \in D} |\varphi(p)|.$$

On the other hand, when α varies over $S_1(0) - \Sigma$, $|q + r_i\alpha| > r_0$, so that

$$\left| \int_{S_1(0)-\Sigma} [\varphi (q + r_1\alpha) - \varphi (q + r_2x)] \, d\alpha \right| \leqslant \varepsilon (r_2 - r_1) \int_{S_1(0)-\Sigma} d\alpha$$

$$\leqslant 4\pi\varepsilon (r_2 - r_1).$$

Now take $\eta > 0$. The difference $r_2 - r_1$ can be chosen independent of the point q, as long as $|q|$ is large enough, since the only condition on $B_{r_2}(q)$ and $B_{r_1}(q)$ is that W lies inside one and outside the other. For the given constant η, take

$$\varepsilon < \frac{\eta}{8\pi (r_2 - r_1)}.$$

Having fixed ε, $\sup\limits_{^\eta r_0} |\varphi|$ is fixed, and we can choose

$$\delta < \frac{\eta}{4 \sup\limits_{B_{r_0}} |\varphi|}.$$

Then, we have

$$\left| \left(\int_{S_{r_1}(q)} - \int_{S_{r_2}(q)} \right) \varphi \, d\alpha \right| < \eta$$

if $|q|$ is large enough. Then, (6.6) shows that[1]

$$\varphi(q) = \frac{1}{4\pi r_2^2} \int_{S_{r_2}(q)} \varphi \, dA + o(1) \quad \text{as} \quad |q| \to \infty. \tag{6.7}$$

A proof similar to the one in § 2 shows that the first term in (6.7) is constant, independent of q. Calling this constant C, we find

$$\varphi(q) = C + o(1) \quad \text{as} \quad |q| \to \infty,$$

so that

$$\lim_{|q| \to \infty} \varphi(q) = C.$$

Thus for the lemma. We now turn to the theorem. Subtracting C from φ and calling the result φ again (a process has no effect on either the hypo-

---

[1] The notation $o(1)$ as $|q| \to \infty$ means some function that goes to zero as $|q| \to \infty$.

theses or the conclusion), we find that we have a function φ satisfying

$$\frac{\partial \varphi}{\partial n} = 0 \quad \text{on} \quad \partial V \tag{6.2}$$

$$\varphi \to 0 \quad \text{as} \quad |q| \to \infty. \tag{6.8}$$

Under these conditions, we show that $\varphi \equiv 0$ in V.

As in the proof of the lemma, let W be the complement of V, and let $B_{r_0}(0)$ be a ball containing W. If $r \geqslant r_0$, Green's theorem gives

$$\int_{V \cap B_r(0)} |\nabla \varphi|^2 \, dq = \int_{S_r(0)} \varphi \varphi_r \, dA,$$

since $\partial \varphi / \partial n = 0$ on ∂V. Once again, let $d\alpha$ denote the element of solid angle, so that $dA = r^2 \, d\alpha$. Then, we find

$$\int_{V \cap B_r(0)} |\nabla \varphi|^2 \, dq = \frac{r^2}{2} \frac{\partial}{\partial r} \int_{S_1(0)} \varphi^2 \, (r\alpha) \, d\alpha.$$

By the lemma and our redefinition of φ, $\int_{S_1(0)} \varphi^2 \, (r\alpha) \, d\alpha$ goes to zero as $r \to \infty$. Therefore,

$$\int_{r_0}^{\infty} \frac{2}{r^2} \int_{V \cap B_r(0)} |\nabla \varphi|^2 \, dq \, dr = - \int_{S_1(0)} \varphi^2 \, (r_0\alpha) \, d\alpha.$$

The left side of this equation cannot be negative, while the right side cannot be positive. The theorem follows from this.

Exercises

6.1   Prove, as stated in the text, that the first term occurring in (6.7) is constant.

6.2   Generalize lemma 6.2 to the following. *Let V be an exterior domain. Let g be a given continuous function on ∂V. Let φ be a velocity potential satisfying the two conditions*

$$\frac{\partial \varphi}{\partial n} = g \quad on \quad \partial V, \tag{6.9}$$

and

$$|\nabla \varphi| \to 0 \quad as \quad |q| \to \infty. \tag{6.10}$$

Then, $\psi(q)$ approaches a constant as $|q| \to \infty$.

6.3   A flow is called *two-dimensional* if there is a rectangular coordinate system such that everything (velocity, pressure, etc.) is independent of z and the velocity w in the z-direction is zero. Prove the result stated in exercise 2 for two-dimensional flows.

6.4   Consider a two-dimensional flow in the exterior domain V consisting of the entire plane slit along the x-axis from 0 to 1. Introduce polar coordinates (r, θ), and define $\varphi (r, \theta) = \log r$. Show that φ is a harmonic function in V satisfying (6.9) (with $g = 0$) and (6.10). Since φ does not satisfy the conclusion of exercise 2, this example illustrates the necessity of the hypothesis made in § 1 that the first derivatives of φ be continuous in the *closure* of V.

6.5   Prove theorem 6.1 for two-dimensional flows.

7   Uniqueness of ideal fluid flows

Theorem 6.1 is remarkable in that it asserts the uniqueness of potential flows without any reference to *initial* conditions. Clearly, a solution of Liouville's equation cannot be ascertained without reference to its initial value. This anomaly in potential flows is due to the fact that in such flows t occurs only as a parameter with respect to which φ is differentiated to determine the pressure. We cannot hope to be so lucky for general ideal fluid flows and must expect an assumption on initial conditions to appear in any theorem about such flows.

Just what initial conditions have to be assumed is not so clear. Given the initial probability distribution $F^0(\xi)$ of chapter one, the initial density, velocity, pressure, etc., of the fluid can be determined from (1.4.1), (1.4.2), (1.4.10), etc. In incompressible fluids, the density is assumed constant, so we can forget about that quantity. For *potential* flows, the pressure is determined completely from Bernouilli's equation if the velocity is given, and Lagrange's theorem 2.3.1, along with the result of the last section, tells us that the velocity is determined once its initial value is given. Consequently, we may hope that even in ideal fluid flows everything is determined once the initial value of the velocity is given.

With these considerations in mind, we consider the following problem. To determine whether or not there can be more than one solution of the equations of ideal fluid flow:

$$s_t + (s \cdot \nabla) s = - \frac{1}{\varrho} \nabla p + Q, \qquad (7.1)$$

$$\nabla \cdot s = 0, \qquad (7.2)$$

in a domain V, where these solutions satisfy the boundary condition

$$n \cdot s = 0 \quad \text{on} \ \partial V, \qquad (7.3)$$

and the initial condition

$$s = s^0 \quad \text{for} \ t = 0. \qquad (7.4)$$

Here, s^0 is a given function with continuous first derivatives in the closure of V satisfying

$$\nabla \cdot s^0 = 0. \qquad (7.5)$$

By a solution of (7.1–4), we mean a function with continuous first derivatives in[5] $[0, T] \times \bar{V}$ for some $T > 0$. If, as usual, s is the vector (u, v, w), we

---

[5] By $\bar{V}$, we mean the closure of V.

use the notation ∇s for the matrix

$$s = \begin{pmatrix} u_x & u_y & u_z \\ v_x & v_y & v_z \\ w_x & w_y & w_z \end{pmatrix}.$$

The *length* of a matrix (denoted by the usual absolute value sign) is defined as the square root of the sum of the squares of the entries in the matrix.

THEOREM 7.1   *Let V be a bounded domain. Then, the problem (7.1–4) has at most one solution.*

Proof   Let s_1 and s_2 be solutions of (7.1–4) with corresponding pressures p_1 and p_2. Define $s = s_1 - s_2$ and $p = p_1 - p_2$. Then, s and p satisfy the equations

$$s_t + (s_1 \cdot \nabla) s + (s \cdot \nabla) s_2 = -\frac{1}{\varrho} \nabla p, \tag{7.6}$$

$$\nabla \cdot s = 0, \tag{7.7}$$

$$n \cdot s = 0 \quad \text{on } \partial V, \tag{7.8}$$

$$s = 0 \quad \text{for} \quad t = 0. \tag{7.9}$$

Multiply (7.6) by s, and integrate the result over V. We find

$$\frac{d}{dt} \int_V |s|^2 \, dq + \int_V s \cdot \nabla s_2 \cdot s \, dq$$

$$= -\frac{1}{q} \int_V s \cdot \nabla p \, dq - \frac{1}{2} \int_V s_1 \cdot \nabla |s|^2 \, dq. \tag{7.10}$$

Because of (7.7) and the equation $\nabla \cdot s_1 = 0$, the right side of (7.10) is

$$-\int_V \nabla \cdot \left(\frac{1}{\varrho} ps + \frac{1}{2} |s|^2 s_1 \right) dq = -\int_{\partial V} n \cdot \left(\frac{1}{\varrho} ps + \frac{1}{2} |s|^2 s_1 \right) dA$$

$$= 0$$

by (7.8). Thus, (7.10) gives

$$\frac{d}{dt} \int_V |s|^2 \, dq = -\int_V s \cdot \nabla s_2 \cdot s \, dq$$

$$\leqslant \sup_V |\nabla s_2| \int_V |s|^2 \, dq. \tag{7.11}$$

Let $c = \sup_V |\nabla s_2|$. An equivalent form of (7.11) is

$$\frac{d}{dt} \left(e^{-ct} \int_V |s|^2 dq \right) \leqslant 0,$$

and this shows that $e^{-ct} \int_V |s|^2 \, dq$ is non-increasing as a function of t.

Since (7.9) shows that $e^{-ct} \int_V |s|^2 \, dq$ is initially zero, we see that

$$e^{-ct} \int_V |s|^2 \, dq \leqslant 0. \tag{7.12}$$

Finally, since the left side of (7.12) can never be negative, we conclude that it is zero. The theorem follows from this.

Exercise

7.1 State and prove a theorem corresponding to 7.1 for exterior domains.

CHAPTER 4

Potential flows in two dimensions

1 Introduction

IN THIS and the next few chapters, we shall be discussing flows in two dimensions. This means (cf. exercise 4.6.3) that there is a rectangular coordinate system in which all quantities that appear are independent of z and the component of the velocity in the z-direction—which we have been calling w— is identically zero. In such a case, the flow in every plane $z =$ constant is the same and what happens in one of them happens in all. This type of flow is approximated by the flow over a long, non-tapered wing, such as that shown in Figure 5. The flow to the left of the dotted lines is surely greatly affected by the presence of the fuselage, while that to the right is affected by so-called "tip effects", but between the dotted lines the flow can often be approximated very well by two-dimensional flow.

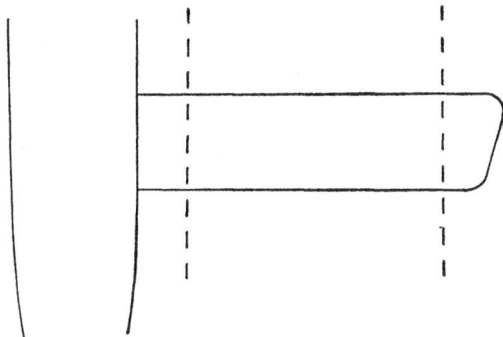

Figure 5

For two-dimensional potential flows, we have a velocity potential $\varphi\,(x, y)$ satisfying

$$\nabla^2\varphi = \varphi_{xx} + \varphi_{yy} = 0. \tag{1.1}$$

It is well known that, corresponding to any solution of (1.1), there is an *analytic* function of the complex variable $x + iy$ of which φ is the real part. Since, in two-dimensional flows, the variable that we have been calling z until now has no role, we are free to use this symbol for $x + iy$. We shall adhere to this notation in the next few chapters, writing

$$z = x + iy.$$

Let f be the analytic function of which φ is the real part, and denote the imaginary part of f by ψ, so that

$$f(z) = \varphi\,(x, y) + i\psi\,(x, y).$$

The function ψ is called the *stream-function* of the flow of which φ is the velocity potential. The function f is often called the *complex velocity potential*.

In addition to depending on z, the complex velocity potential may depend on t. If f is independent of t, the flow is called *steady*. In steady flow, a curve that moves with the fluid is often called a *streamline*.

If f is analytic, the Cauchy-Riemann equations apply to its real and imaginary parts. We have, therefore,

$$\begin{aligned}
u &= \varphi_x = \psi_y, \\
v &= \varphi_y = -\psi_x.
\end{aligned} \tag{1.2}$$

This fact can be used to prove

THEOREM 1.1   *In steady flow, the stream function is constant along streamlines.*

Proof   Let $x = x(t)$, $y = y(t)$ be the coordinates of any point that moves with the fluid. Since the flow is steady,

$$\frac{d}{dt}\,\psi\,(x(t), y(t)) = \psi_x\dot{x} + \psi_y\dot{y} = \psi_x u + \psi_y v = 0,$$

by (1.2). This proves the theorem.

A related result, valid for any potential flow, steady or not, is

THEOREM 1.2   *The stream function may be taken to be constant along a fixed boundary of a flow.*

Proof   If we have a fixed boundary of which n denotes the normal and τ the tangential direction, we have, according to the Cauchy-Riemann equations,

$$\frac{\partial\varphi}{\partial n} = \frac{\partial\psi}{\partial\tau}.$$

Along a fixed boundary, however, $\partial\varphi/\partial n = 0$. Therefore, on a boundary, ψ is a function of t alone. Since subtraction of a function of t from the stream function has no effect on the flow, theorem 1.2 follows.

Of course, along two disconnected fixed boundaries, the constant values of ψ may be different.

2   Examples of two-dimensional potential flows

The results of the foregoing section show that, associated with every analytic function f of the complex variable z is a two-dimensional, potential flow. The boundaries of the flow are defined either by the equation $\partial\varphi/\partial n = 0$ or by $\psi = $ constant. That is, any curve $\psi = $ constant can be replaced by a fixed wall since there is never any flow through such a curve. In this section, we look at a number of examples of steady, two-dimensional flows obtained by examination of analytic functions.

(1)   One of the simplest analytic functions is a power of z. Let the complex velocity potential be

$$f(z) = Az^n,$$

where A is a positive constant. In polar coordinates $z = r\,e^{i\theta}$, we have

so that
$$f(z) = Ar^n\,e^{in\theta} \qquad (2.1)$$

$$\varphi = Ar^n \cos n\theta,$$

$$\psi = Ar^n \sin n\theta.$$

(a)   If $n = 1$, we have $\varphi = Ax$, $\psi = Ay$. Thus, the streamlines are all lines parallel to the x-axis, and the flow is the uniform one parallel to the x-axis with speed A.

(b)   Let $n = 2$. Then,

$$\psi = Ar^2 \sin 2\theta = 2Axy.$$

The streamlines here are hyperbolas. The streamline corresponding to $\psi = 0$ consists of the coordinate axes. Replacing the axes $x > 0, y = 0$ and $y > 0$, $x = 0$ by fixed walls, we obtain the flow in the region shown in Figure 6.

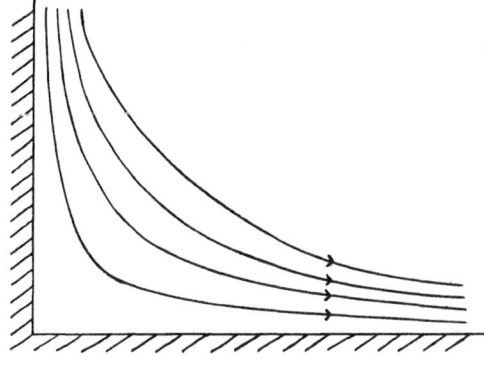

Figure 6

An obvious calculation shows that the components of the velocity at any point are

$$u = 2Ax$$

$$v = -2Ay.$$

(c) The flow between two walls making an angle α with each other has a complex velocity potential of the form (2.1) for any α. Let one of the walls be the positive x-axis, so that it has the equation $\theta = 0$. Let the other be the line $\theta = \alpha$. The streamlines associated with (2.1) have the equation

$$\psi = Ar^n \sin n\theta = \text{constant}.$$

Since $\sin n\theta = 0$ when $\theta = 0$ and when $\theta = \pi/n$, we see that the two rays $\theta = 0$ and $\theta = \pi/n$ correspond to the same streamline. Set $n = \pi/\alpha$. Then, the rays $\theta = 0$ and $\theta = \alpha$ lie on the same streamline, which we may consider to be a rigid wall. The associated velocities are

$$\varphi_r = -\frac{\pi A}{\alpha} r^{(\pi/\alpha)-1} \cos \frac{\pi\theta}{\alpha},$$

$$\varphi_\theta = \frac{\pi A}{\alpha} r^{\pi/\alpha} \sin \frac{\pi\theta}{\alpha}.$$

Clearly there is an infinite velocity at the origin if $\alpha > \pi$. Presumably in this case the hypotheses of ideal fluid flow break down in a neighborhood of the origin. What will be much more important to us later is the mere fact that the velocity is infinite for flow past a convex corner and finite otherwise. We state this as

THEOREM 2.1   *There exists a two-dimensional potential flow past the wedge $0 < \theta < \alpha$ having an infinite velocity at the edge of the wedge if $\alpha > \pi$ and having a finite velocity there if $\alpha \leqslant \pi$.*

(2) A point at which fluid is being created at a constant rate is called a *source*. (If the rate is negative, it is called a *sink*.) At a source, fluid will flow outward in all directions along rays through the source. If μ denotes the volume of fluid being created per unit time (the *strength* of the source), then the radial velocity at a distance r from the source is

$$R = \frac{\mu}{2\pi r};$$

the angular velocity is zero.

Setting

$$\varphi_r = \frac{\mu}{2\pi r},$$

$$\varphi_\theta = 0,$$

we find that, except for a constant,

$$\varphi = \frac{\mu}{2\pi} \log r.$$

This is a solution of Laplace's equation (except, of course, at $r = 0$), as a direct computation shows. Since

$$f(z) = \frac{\mu}{2\pi} \log z = \frac{\mu}{2\pi} (\log |z| + i \arg z) = \frac{\mu}{2\pi} (\log r + i\theta)$$

is an analytic function of z, we infer that $\log z$ is the corresponding complex velocity potential. The stream function is

$$\psi = \frac{\mu}{2\pi} \theta = \frac{\mu}{2\pi} \arctan \frac{y}{x}.$$

(3) Of course, any two solutions of (1.1) may be superposed to obtain another. Consider, then, two sources of equal strength at the points $z = \pm h$. The corresponding stream functions are,

$$\psi^{\pm} = \frac{\mu}{2\pi} \arctan \frac{y}{x \mp h}.$$

We conclude, then, that the function

$$\psi = \frac{\mu}{2\pi} \left[\arctan \frac{y}{x + h} + \arctan \frac{y}{x - h} \right] \tag{2.2}$$

is also the stream function of a flow.

When $x = 0$, then $\psi = 0$; thus, the y-axis may be considered to be a rigid barrier. The rate of flow across any circle around the point $x = h$ is equal to the rate of flow due to the source at $x = h$. The reason is that the function $\mu/2\pi \arctan y/(x + h) = \psi^-$ is harmonic at $x = h$. Therefore, if φ^- denotes the corresponding velocity potential, equation (3.2.1) shows that the rate of flow across the circle S_r of radius r about $z = h$ is

$$\frac{1}{2\pi} \int_0^{2\pi} \frac{\partial \psi}{\partial n} d\theta = 0.$$

Thus, (2.2) can be looked upon as the stream function associated with a source in the presence of a rigid wall. The streamlines are sketched in Figure 7. They are all rectangular hyperbolas passing through the source.

(4) Another interesting case is obtained if one considers, not two sources, but a source and a sink of equal strength, placed at the points $z = \pm h$, say. From what we now know, we can write down the associated stream function

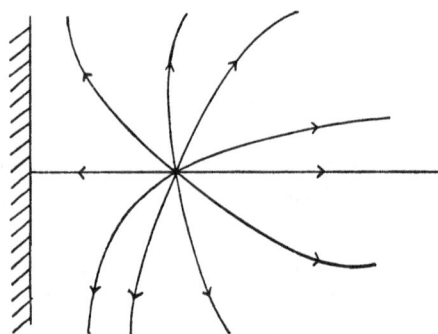

Figure 7

immediately. It is

$$\psi = \frac{\mu}{2\pi}\left[\arctan\frac{y}{x+h} - \arctan\frac{y}{x-h}\right]. \tag{2.3}$$

If a uniform flow $\psi = Uy$ is superposed on this, one can obtain the stream function of a flow, uniform at infinity, over an ovular body.

The simplest case occurs when h is allowed to approach zero, μ approaching infinity at the same time in such a way that the product remains constant. The result is called a *doublet*. Set

$$2\mu h = m.$$

As $h \to 0$, the stream function (2.3) approaches

$$= \frac{-m}{2\pi r}\sin\theta, \tag{2.4}$$

and this is also a stream function. (Any limit of solutions of (1.1) in a fixed domain is also a solution.) If we add to (2.4) the stream function associated with a uniform flow, we obtain

$$\psi = Uy - \frac{2\pi}{m}\frac{\sin\theta}{r} = Uy - \frac{m}{2\pi}\frac{y}{r^2}. \tag{2.5}$$

The streamline $\psi = 0$ consists of two parts, the line $y = 0$, and the circle

$$r^2 = \frac{m}{2\pi U}. \tag{2.6}$$

Thus, (2.5) can be looked upon as the flow past the circle (2.6) having the constant velocity $(U, 0)$ at infinity. The corresponding velocity potential is

$$\varphi = Ux + \frac{m}{2\pi}\frac{x}{r^2},$$

and the complex velocity potential is

$$f = Uz + \frac{m}{2\pi z}. \tag{2.7}$$

The flow pattern is sketched roughly in Figure 8.

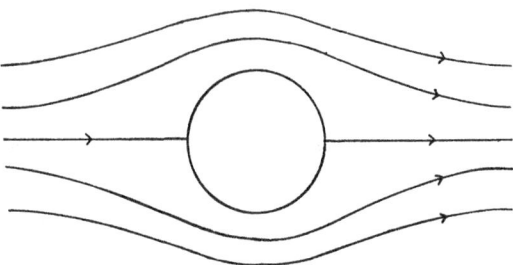

Figure 8

The solution (2.7) is extremely important, and we shall return to it in the next section.

3 Existence of potential flows

We saw in § 3.6 that there cannot be more than one flow satisfying certain reasonable hypotheses. But the question remains if in general there are any flows at all in a given domain. For two-dimensional flows past a rigid body, this question can now be settled.

We prove the existence of a velocity potential in certain very general exterior domains V. Because of the generality of the domains allowed, a few remarks about the theorem are necessary to understand it. First, we note that the requirement in chapter three that the velocity potential be continuously differentiable in the closure of the domain must be relaxed, for it can be proved that unless ∂V is smooth, it is impossible in general for there to be a non-trivial velocity potential in V whose derivatives are even bounded near ∂V. Examples of this phenomenon have already been given by § 3 and exercise 3.6.4.

A consequence is that under the conditions of the theorem below, the velocity potential is *not* unique. To prove this, let V_1 be the complex plane slit along the real axis from $z = 0$ to $z = 1$. The theorem will tell us that this is a harmonic function φ_1 in V_1 with zero normal derivative on ∂V_1 such that

$$\nabla \varphi_1 \to (1, 0)$$

(say) as $z \to \infty$. Let φ be the velocity potential of exercise 3.6.4. Then $\varphi_1 + \varphi$ is another velocity potential satisfying the same conditions as φ_1. This shows that the velocity potential need not be unique[1].

Next, we have to say what we mean by a velocity potential in a domain of the type we are discussing. For "reasonable" domains, the boundary condition

$$\frac{\partial \varphi}{\partial n} = 0 \tag{3.1}$$

that we have been using all along makes perfect sense, and can even be interpreted as saying that the fluid never penetrates the boundary of V. But for domains of the type we consider here, V need not even have a normal vector at each point of its boundary. What then shall we make of (3.1)? The answer lies in theorem 1.2. Condition (3.1) can be replaced by the condition that ψ is constant on ∂V, or better, by the weak form

$$\psi(x, y) \to \text{constant} \quad \text{as} \quad (x, y) \to \partial V.$$

With this hint, we can define precisely what properties we might hope a velocity potential to have. We say that $f(z)$ is a *weak complex velocity potential* for the domain V if $f(z)$ is analytic in V and

$$\text{Im } f(z) \to \text{constant} \quad \text{as} \quad z \to \partial V \quad \text{from within } V.$$

Next, a word about the point at infinity. The last example of § 2 shows that for exterior domains it is too much to expect that a complex velocity potential be analytic at infinity. What we generally want is for the velocity to be given at infinity (corresponding to the physical situation of air blowing over a fixed object). If f is a complex velocity potential for the exterior domain V, let

$$f = \varphi + i\psi.$$

The derivative of f is given by

$$f' = \varphi_x + i\psi_x = \varphi_x - i\varphi_y,$$

by the Cauchy–Riemann equations. This means that if the velocity is given at infinity, say as the vector (u_∞, v_∞), we must have

$$f'(z) \to u_\infty - iv_\infty \quad \text{as} \quad z \to \infty.$$

Therefore, we expect f to look like

$$f(z) = (u_\infty - iv_\infty) z + (\text{bounded terms})$$

---

[1] On the other hand, it can be proved that if ∂V is smooth, then there is a velocity potential in V whose derivatives are continuous in $\overline{V}$. Exercise 3.6.5 shows that such a potential is unique.

as $z \to \infty$. This means that $f(z)$ has a simple pole at infinity, and that is precisely what we prove happens in general.

We can now state

THEOREM 3.1 *Let V be an exterior domain with a connected boundary and at least two boundary points. Then there is a weak complex velocity potential for V with a simple pole at infinity whose associated velocity is prescribed at infinity.*

To prove the theorem, we begin with a lemma, following which it will be possible simply to write down the complex velocity potential whose existence is asserted.

LEMMA 3.2 *Let V be an exterior domain with a connected boundary and at least two boundary points. Then, there is a function k, analytic in V except for a simple pole at infinity, mapping V conformally onto the exterior of the unit disk, and satisfying*

$$k(\infty) = \infty,$$

$$k'(\infty) > 0. \tag{3.2}$$

Proof Let z_0 be any point on ∂V. Define

$$\zeta = g(z) = \frac{1}{z - z_0}. \tag{3.3}$$

Let V' be the image of V under g. We prove that V' is simply connected. Let γ' be a simple closed curve in V', and let γ be its pre-image under the map (3.3). Suppose there were a point ζ_1 on $\partial V'$ and in the interior of γ'. Define z_1 by the equation $\zeta_1 = g(z_1)$. γ divides the z-plane into two connected components. Moreover, z_0 and z_1 are both points of ∂V, which is also connected. Therefore, z_0 and z_1 lie on the same side of γ. Therefore, the images of z_0 and z_1 must both lie inside or outside γ'. But this is patently false, since ζ_1 lies inside γ', while the image of z_0 is ∞, which lies outside γ'. Thus every point inside γ' is a point of V', and V' is simply connected.

According to the Riemann mapping theorem[1], since V' is simply connected there is a function

$$\omega = h(\zeta),$$

analytic in V', mapping V' onto the disk $|\omega| < 1$, such that

$$h(0) = 0,$$

$$h'(0) > 0.$$

_____

[1] See, *e.g.*, Zeev Nehari, Conformal mapping. McGraw-Hill, 1952.

Finally, let

$$w = j(\omega) = \frac{1}{\omega}.$$

j is analytic in $|\omega| < 1$ punctured at $\omega = 0$, and it maps this domain onto $|w| > 1$ punctured at infinity.

Let k be the composition of g, h, and j:

$$w = k(z) = j\left(h\left(g(z)\right)\right).$$

$k(z)$ is analytic in $V - \{\infty\}$, since g is analytic in V, h in V', and j in the punctured disk. $k(z)$ also has a simple pole at infinity. Recall that $k(z)$ is said to have a simple pole at infinity if $k\,(1/z)$ has a simple pole at zero. Now,

$$k\left(\frac{1}{z}\right) = \frac{1}{h\,[z/(1 - z_0 z)]}.$$

Both h and the function $z/(1 - z_0 z)$ are analytic at zero. Moreover, since $h(0) = 0$, the Taylor series for $h\,(z/1 - z_0 z)$ near zero looks like

$$h'(0)\,z + \cdots$$

Therefore,

$$zk\left(\frac{1}{z}\right) = \frac{1}{h'(0)} + \cdots,$$

the dots here representing powers of z of order greater than or equal to one. Thus, $k(z)$ has a simple pole at infinity.

Clearly, $k(\infty) = \infty$. To prove (3.2), consider

$$k'(z) = j'(\omega)\,h'(\zeta)\,g'(z) = \frac{h'(\zeta)}{(z - z_0)^2\,\omega^2}.$$

The limit of the quantity $(z - z_0)\,\omega$ occurring here can be found by using l'Hôspital's rule. The result is

$$\lim_{z \to \infty}(z - z_0)\,\omega = \lim_{z \to \infty}(z - z_0)\,h\left(\frac{1}{z - z_0}\right) = \lim_{z \to \infty}h'\left(\frac{1}{z - z_0}\right) = h'(0).$$

Therefore,

$$k'(\infty) = \frac{1}{h'(0)},$$

which is positive. This completes the proof of the lemma.

To prove the theorem, let f be a complex velocity potential associated with the domain $|w| > 1$. According to example (4) of the last section, we can take

$$f(w) = U^*\left(w + \frac{1}{w}\right), \tag{3.2}$$

where U^* is any real number. We prove that U^* can be so chosen that

$$F(z) = f(k(z))$$

is a weak complex velocity potential satisfying all the conditions of theorem 3.1. It is clear that F is analytic in V. Moreover, as $z \to \partial V$, $|w| = |k(z)| \to 1$, and (3.2) shows that Im $f(w) \to 0$. Therefore, Im $F(z) \to 0$, and ∂V is a streamline.

Next, we have

$$F'(\infty) = f'(\infty) k'(\infty) = U^*k'(\infty). \tag{3.3}$$

The velocity of the flow is given at infinity. By rotating the coordinate system if necessary, the given velocity at infinity may be made to have zero component in the y-direction, so that it has the form $(U, 0)$. Then, as we saw in the discussion preceding theorem 3.1, what we want is

$$F'(\infty) = U.$$

Now, $k'(\infty) > 0$, and U^* is free. Taking

$$U^* = \frac{U}{k'(\infty)},$$

we achieve the desired result.

We have actually proved something more specific than was stated in theorem 3.1. The actual result is

THEOREM 3.3   *Let V be an exterior domain with a connected boundary and at least two boundary points. Let k be an analytic function in V, having a simple pole at infinity, mapping V conformally onto the exterior of the unit disk, and satisfying $k'(\infty) > 0$. Then, a weak complex velocity potential for V is*

$$F(z) = U\left[k(z) + \frac{1}{k(z)}\right],$$

where U is real. The velocity of the flow at infinity is parallel to the real axis and has magnitude $Uk'(\infty)$.

Exercises

3.1 In proving theorem 3.1, we mapped V onto the exterior of the unit disk and then used the known complex velocity potential for the latter domain to derive the potential for V. This is the natural thing to do if the complement of V is bounded. For flows in a "channel", however, it is not at all natural. Let us call V a *channel* if it is the domain between two non-intersecting curves going to infinity, asymptotic to parallel straight lines there. (See Figure 9.) In figure 9, we conceive of the flow as moving from left to right. The left of the figure is called the *upstream* direction, the right *downstream*.

(a)   Show that any channel can be mapped conformally onto a channel whose boundary is two straight lines parallel to the real axis.

(b)   Show that a potential flow in a channel bounded by two straight lines is uniform flow.

(c)   Using the results of (a) and (b), show that there exists a weak velocity potential in any channel with the velocity at infinity *upstream* given parallel to the walls of the channel there.

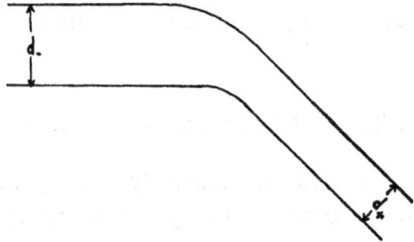

Figure 9

(d)   If, as indicated in figure 9, the distance between the walls upstream and down are d_- and d_+, respectively, show that the velocity at infinity downstream is

$$U \frac{d_-}{d_+}$$

if U is the velocity at infinity upstream.

This exercise shows that one can compute quite detailed information about certain potential flows, information that can be checked against experiment to draw conclusions about the validity of the hypotheses that underlie potential flow.

3.2   Show that the function

$$k(z) = 2z - 1 + 2\sqrt{z(z - 1)}$$

maps the plane slit along the real axis from $z = 0$ to $z = 1$ onto the exterior of the unit disk. Use this fact and the result of exercise 3.6.4 to derive another flow past the disk, different from the one derived in § 2.

3.3   Show there is a weak complex velocity potential in the half-space $y > 0$ with velocity $(\cos\alpha, -\sin\alpha)$ at infinity for any α in the interval $0 \leqslant \alpha \leqslant \pi/2$. For what values of α can the adjective "weak" be dropped?

4   Examples of two-dimensional potential flows (continued)

With the aid of theorem 3.3, we can construct potential flows at will, given any function mapping a domain onto the exterior of the disk. We give some examples of such constructions here.

(5)   *The flat plate*   It is easy to show that the transformation

$$k(z) = z + \sqrt{z^2 - e^{-2i\alpha}}$$

maps the complex plane slit along the line segment with length two, centered at zero, and making an angle α with the negative real axis onto the exterior of the unit disk. Since $k'(\infty) = 2$, a weak potential flow associated with the slit domain having velocity $(U, 0)$ at infinity is the flow described by the complex velocity potential

$$F(z) = \frac{U}{2} [(1 + e^{2i\alpha}) z + (1 - e^{2i\alpha}) \sqrt{z^2 - e^{-2i\alpha}}]. \qquad (4.1)$$

A point in a flow where the velocity is zero is called a *stagnation point*. For potential flows, a stagnation point is any point where $F'(z)$ is zero. The flow described by (4.1) has stagnation points at

$$z = \pm e^{-i\alpha} \sqrt{\frac{1 + \cos 2\alpha}{2}},$$

two points lying on the slit. Knowing the stagnation points, it is not hard to sketch roughly the streamlines of the flow. Such sketches are shown in Figure 10 for three different values of α.

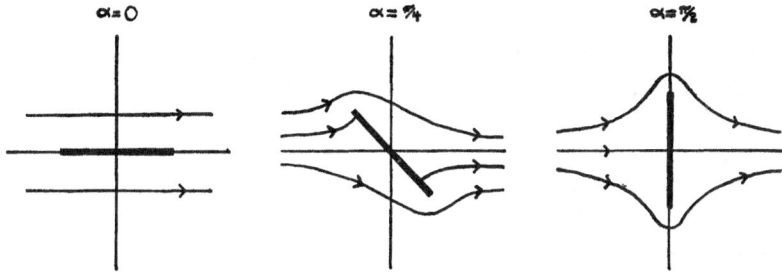

Figure 10

(6) *Joukowski airfoils*   Thus far, we have had no examples that resemble those of the most important application, the wing of an airplane. The first such class of examples was discussed by Joukowski[1]. We describe his airfoils now.

It is convenient in what follows to refer, not to the function $k(z)$ that maps a domain onto the exterior of a disk, but to its inverse. We write $j(w) = k^{-1}(w)$, so that $j(w)$ maps the disk $|w| > 1$ onto the domain V of the fluid.

To an observer stationed at infinity, an airfoil and a circle are the same; in fact, the presence of neither can be felt at all. Therefore, for any shape airfoil, the function $j(w)$ should look, near infinity, like the mapping function for

---

[1] N. Joukowski, Über die Konturen der Tragflächen der Drachenflieger. Zeit. für Math. (1910).

the disk. That is, $j(w)$ should resemble a constant times w. Dropping the constant for now, this means that $j(w) - w$ approaches a limit at infinity. For the maps defined by lemma 3.2, $j(w) - w$ is actually analytic at infinity, so that

$$j(w) = w + a_0 + \frac{a_1}{w} + \frac{a_2}{w^2} + \cdots$$

in a neighborhood of infinity.

The constant a_0 is irrelevant: it can be eliminated by translation of the disk in the w-plane. Thus, we may take $a_0 = 0$. Also, rotating the disk about its center amounts to multiplication of a_1 by a number of absolute value one. Therefore, we may assume a_1 is non-negative, and write $a_1 = c^2 \geqslant 0$. With these normalizations, the simplest transformation of the above type is obtained by setting $a_i = 0$ for $i \geqslant 2$. The result is called the *Joukowski transformation*:

$$j(w) = w + \frac{c^2}{w}. \tag{4.2}$$

We observe what the Joukowski transformation does to various circles passing through the point $w = c$. First, the circle $|w| = c$ is transformed into the straight line segment with parametric equation $z = 2c \cos \theta, 0 \leqslant \theta \leqslant 2\pi$. Thus, the flows of part (5) are special cases of the flows obtained by means of the Joukowski transformation.

Now consider a circle centered at a point of the imaginary axis, as shown in Figure 11.

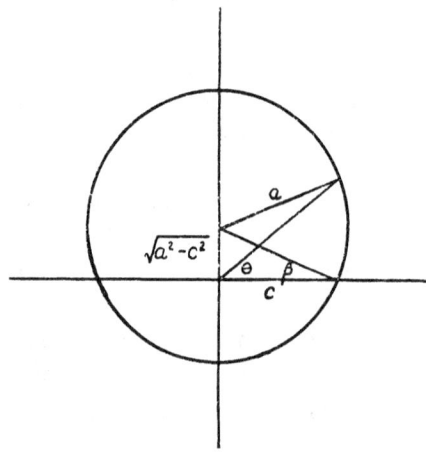

Figure 11

Writing $w = r\, e^{i\theta}$ and $z = j(w) = x + iy$, we find

$$x = \left(r + \frac{c^2}{r}\right) \cos\theta,$$

$$y = \left(r - \frac{c^2}{r}\right) \sin\theta.$$

This means that

$$x^2 \sin^2\theta - y^2 \cos^2\theta = 4c^2 \sin^2\theta \cos^2\theta. \tag{4.3}$$

But the cosine law gives

$$a^2 = a^2 - c^2 + r^2 - 2r\sqrt{a^2 - c^2}\,\sin\theta,$$

while clearly

$$a = c \sec\beta.$$

Thus, we find

$$r^2 - c^2 = 2cr \tan\beta \sin\theta,$$

or

$$y = \left(r - \frac{c^2}{r}\right)\sin\theta = 2c \tan\beta \sin^2\theta. \tag{4.4}$$

Eliminating θ between (4.3) and (4.4), we see that the image in the z-plane of the circle sketched in figure 11 has the equation

$$x^2 + (y + 2c \cot 2\beta)^2 = (2c \csc 2\beta)^2. \tag{4.5}$$

This is the equation of a circle. But (4.4) shows that $y \geqslant 0$. Consequently, the image of the circle of Figure 11 is that portion of the circle (4.5) that lies above the x-axis. Translating the coordinate system in the w-plane so that the origin is at the center of the circle of Figure 11, and using theorem 3.3 we can find the velocity potential of the flow past any circular arc like that shown in Figure 12.

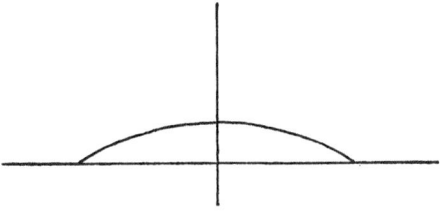

Figure 12

One can do better than this. The image under the Joukowski transformation of a circle in the w-plane with center on the negative real axis passing through $w = c$ is a symmetrical airfoil looking like the sketch in Figure 13.

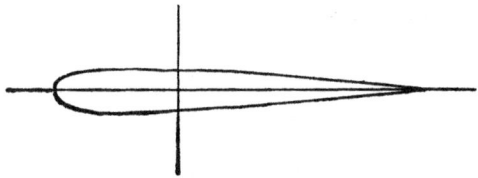

Figure 13

Finally, if the circle in the w-plane is taken to have its center in the second quadrant, then the Joukowski transformation maps it into an airfoil shape like that shown in Figure 14[1].

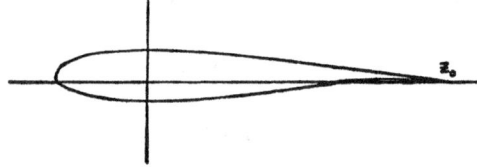

Figure 14

Exercises

4.1   Find the velocity potential of the flow past the circular arc of figure 12.

4.2   Choose a circle in the w-plane having center on the negative real axis, find its image under the Joukowski transformation, and sketch the symmetrical Joukowski airfoil thereby obtained.

---

[1] For more details of the Joukowski transformation as well as extensions of it, see the book by H. Glauert, The elements of aerofoil and airscrew theory. Cambridge University Press, 1947. Figures 13 and 14 are taken from Glauert's book.

CHAPTER 5

D'Alemberts paradox and early attempts at its resolution

1 Introduction

THE SIMPLE hypothesis of potential flow has carried us very far. Farther, in fact, than a student on his first exposure is probably prepared to believe. And yet, the fact that we can describe flows about such complicated shapes as that of Figure 14, and the additional fact that these flows in their gross structure agree with experiment (as they do!), is nothing other than extraordinary.

But our luck can only be stretched so far. In many crucial ways, the flows that we have found have features in sharp disagreement with experiment. Perhaps the most striking of these is d'Alembert's paradox: the net force on any body in potential flow is zero. In this chapter, we discuss this paradox and an attempt, a century ago, to describe another model of the flow past a body, one that is different from what we have discussed so far and one that is not subject to this paradox.

First, a remark about the word "paradox". I do not know who first used it in this context, but it should be realized that the word is being used in a sense that is distinct from the usual one. The barber who shaves everyone in town who does not shave himself is a paradox; d'Alembert's paradox is a theorem. That is, it is correct within the framework of the assumptions made. If the conclusion is not in accord with ordinary experience, then there must be something wrong with the assumptions. But notice that the fact that the result is called a paradox is evidence that early workers in the field had great confidence in potential flows as descriptive of real ones. This in itself shows how successful this model of real flows is in describing their overall features.

2 The d'Alembert paradox

Consider a body with a plane of symmetry orthogonal to the direction of the velocity at infinity. (See Figure 15.) It is not terribly hard to show that the velocity potential of the flow past such a body is symmetric, so that if the plane of symmetry is taken as the y-axis,

$$\varphi(x, y) = \varphi(-x, y). \tag{2.1}$$

We have already seen examples of this in the potential flow past a circle and a vertical plate (see Figures 8 and 10c). Now, the pressure in the flow is given by Bernoulli's equation

$$\frac{1}{\rho} p = \Omega - \frac{1}{2} |\nabla \varphi|^2.$$

If the potential Ω is symmetric, (2.1) shows that the pressure is too.

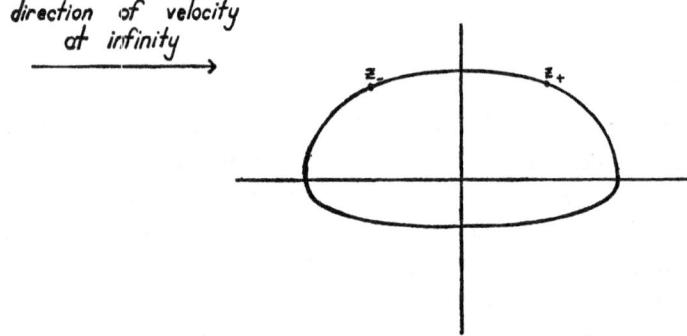

direction of velocity at infinity

Figure 15

The force exerted by the fluid on the body is given by (1.7.6). In any ideal fluid flow, the stress tensor τ is just

$$\tau = pI,$$

where I is the identity matrix. Thus, if n is the exterior normal to the domain of the fluid at any point of its boundary, we have that the force on the body is

$$F = - \int_{\partial V} pn \, d\lambda \qquad (2.2)$$

where λ denotes arclength[1] along ∂V. At any two symmetric points z_+ and z_- on ∂V (see Figure 15) the pressure is the same, while $n(z_+) = -n(z_-)$. Thus, the integral (2.2) vanishes, and the force is zero.

That this is the case for a symmetric body like that of Figure 15 is not so surprising. What is a surprise is that the same conclusion holds even if the body is not symmetrical. In fact, we have

THEOREM 2.1   (d'Alembert's paradox.) *In steady potential flow, the net force on a bounded body is zero.*

It is possible to give a complete proof of this theorem, based on properties of solutions of Laplace's equation[2]. However, the proof is not particularly

---

[1] The area element dA of (1.7.6) is replaced by the element of arclength $d\lambda$ since the flow is two-dimensional.

[2] See, *e.g.*, Garrett Birkhoff, Hydrodynamics, a study in logic, fact, and similitude. Dover, 1950.

enlightening and instead, we give an heuristic argument. That the argument is only heuristic and not rigorous can be seen from the fact that it apparently applies also to flows past unbounded obstacles in a flow while, as we shall see later on in this chapter, such flows are *not* in general subject to d'Alembert's paradox.

Suppose, then, we have a steady flow past a bounded obstacle B. Let $\varphi(x, y)$ be a velocity potential of the flow. A particle moves with the fluid if it satisfies the differential equations

$$\left.\begin{aligned} \dot{x} &= \varphi_x(x, y), \\ \dot{y} &= \varphi_y(x, y). \end{aligned}\right\} \tag{2.2}$$

Now, suppose we let time run backwards. Since the flow is steady, $\varphi(x, y)$ is still a velocity potential of the flow, the only difference between the flows when time goes forward and when it runs backward being that the differential equations (2.2) must be replaced by

$$\left.\begin{aligned} -\dot{x} &= \varphi_x(x, y), \\ -\dot{y} &= \varphi_y(x, y). \end{aligned}\right\}$$

Thus, allowing time to run backwards is the same as reversing the direction of the velocity vector at each point. What is important about this is the fact that the flow with the velocity vector at infinity reversed is obtained by replacing φ by $-\varphi$.

One can also see this directly. If φ^+ is the velocity potential of the flow past an object B with the velocity u^+ at infinity, then the function $\varphi^- = -\varphi^+$ is also a potential, satisfies the same boundary conditions as does φ^+ at the boundary of B, and is associated with the velocity $-u^+$ at infinity.

Let φ be a velocity potential, and consider simultaneously the two potentials $\pm \varphi$. The associated pressures are

$$\frac{1}{\varrho} p_\pm = \Omega - \frac{1}{2} |\pm \nabla\varphi|^2 = \Omega - \frac{1}{2} |\nabla\varphi|^2.$$

We conclude from this that the pressure distribution of a flow coming, say, from the left is identical with the pressure distribution of the flow coming from the right. If, then, there were a net force associated with the left-hand flow, the force due to the right-hand flow would be identical to it, and, in particular, the two forces would have the same direction. Consequently, if a potential flow past a body produced a drag force, the flow in the opposite direction would produce a *negative* drag force: the body would tend to move against the current. This physically unacceptable conclusion completes our argument for d'Alembert's paradox.

3 Cavity flows

A different type of potential flow from the ones that we have been considering
was considered by Helmholtz in 1868 and applied to d'Alembert's paradox
by Kirchhoff in 1869. As an example of this type of flow, we consider the
problem of a two-dimensional flow broadside to a flat plate. Such a flow was
already considered in chapter five; it is sketched in the third part of Figure 10. The associated complex velocity potential is

$$f(z) = \sqrt{z^2 + 1}, \tag{3.1}$$

except for a multiplicative constant.

The complex velocity potential (3.1) has three properties: first, it is analytic
off the plate; second, its imaginary part is constant on the plate; third, its
derivative is given at infinity. If we wish to create a flow different from the
one described by (3.1), we must be prepared to give up one or more of these
properties. The first cannot be given up; it is this property that makes the
real part off a velocity potential and so relates the whole question to fluid
mechanics.

It is possible to give up the second condition. It says there is no flow
through the plate. To give this up, then, would mean that we would be changing our considerations to the study of flow past porous bodies, and we should
still not have resolved any of the paradoxes associated with a non-porous
plate.

Thus, we must be prepared to weaken or change the condition that the
velocity of the flow is given at infinity. Suppose we require that $f'(z)$ be given
at infinity *upstream*, but that downstream of the plate there be a wake, as
sketched in Figure 16. In the wake, we require that the fluid be motionless.
In particular, this implies that the fluid in the wake has a potential associated
with it (the function $\varphi \equiv 0$) and that there is no flow through the back of the
plate.

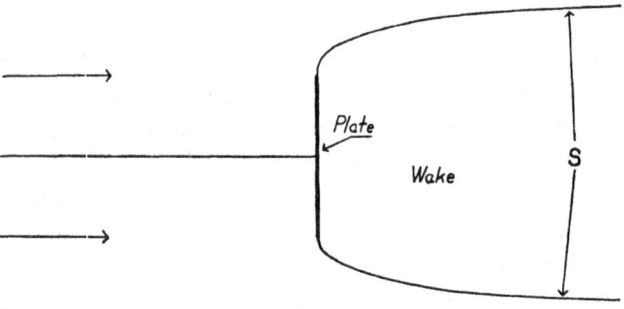

Figure 16

The pressure must be continuous across the curve separating the wake from the rest of the flow. If it weren't, the pressure difference would cause the separating curve to move, and this is impossible in steady flow. But in the wake, the pressure is constant, as Bernoulli's equation shows. Thus, we require that the pressure be constant on the curve S separating the wake from the rest of the flow. Another application of Bernoulli's equation then shows that we must require the speed to be constant on S.

There is another condition to be satisfied on S. If there were flow through S, it would be impossible to satisfy the condition that the velocity be zero in the wake. Therefore, we require that S be a streamline.

What we shall look for, then, is a function $f(z)$, analytic outside of the wake, with imaginary part constant on the plate, and with the further property that on a streamline S through the ends of the plate, the speed, $|f'(z)|$, is constant. $f(z)$ will then be a complex velocity potential of a flow past a plate with a wake.

It turns out that there are many such flows. To specify one, we require that the streamline S extend to infinity, and that the speed on S be the same as the speed of the flow at infinity *upstream*. This still does not determine the flow completely. A certain additional mathematical condition will be imposed in its appropriate place later on.

Assuming for now that it exists, let $f(z)$ be the complex velocity potential associated with the flow. Write

$$\zeta = f(z),$$

and set

$$w = \frac{d\zeta}{dz}.$$

Because of the physical interpretation of the complex velocity potential f, we have

$$w = u - iv,$$

where (u, v) is the velocity vector attached to the point z of the flow.

By our usual assumptions, for each point z of the fluid, there is a corresponding velocity w. Thus, we have a mapping of the domain V exterior to the wake onto a domain H in the w-plane. The w-plane is usually called the *hodograph plane* and H the *hodograph* of V. Once again, the hodograph is the image of the domain of the fluid under the transformation $w = f'(z)$ where $f(z)$ is the complex velocity potential.

We assume that the mapping of V onto H is one-to-one. This is the hypothesis that I referred to earlier. There is no physical reason that I know of for it to be valid, but the assumption allows us to determine the function f. An assumption that the map $w = f'(z)$ is bivalent or trivalent would also produce a flow.

It is not hard to see what the hodograph looks like. The boundary of H is, as usual, the image of the boundary of V. For points on the plate, the imaginary part of f is constant. Thus, for such points, $v = 0$, and the image of such points must lie on the real axis in the hodograph plane.

It is convenient to situate the plate along the *real* axis in the z-plane with the center of the plate at the point $z = 0$. Then the velocity at the point $z = 0$ is zero. Therefore, the image of the plate in the hodograph is a line segment on the real axis with center at $w = 0$.

On the streamlines S, we are assuming that the speed is a constant, so that

$$|w| = \sigma = \text{constant}$$

on S. Thus, the boundary of the hodograph must be the semicircle shown in Figure 17, and the hodograph itself must be either the interior or the exterior of this semicircle. It will turn out as a consequence of our calculations that the velocity of this flow is continuous on ∂V. It follows from this that the hodograph is the interior of the semicircle. For, by the maximum principle, the speed cannot take on a maximum at any point in the fluid itself, while it equals σ on S and is finite on the plate. Thus, the speed is bounded, as is the hodograph.

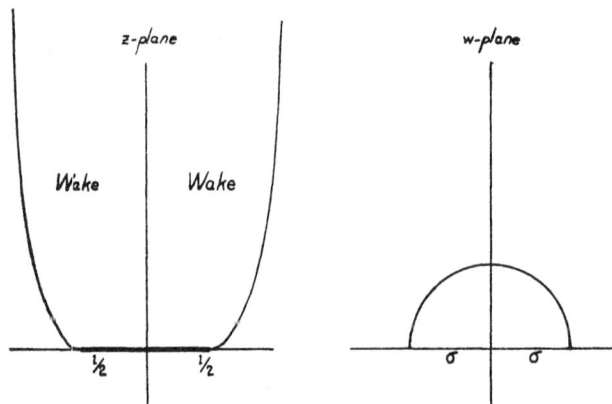

Figure 17

We now look at the variable $\zeta = f(z)$. The image of V in the ζ-plane can easily be seen to be the entire ζ-plane, slit along half of the real ζ-axis[1]. By the addition of a constant to f, we can always make the image the ζ-plane slit along the positive real axis $\zeta > 0$.

Let

$$s = -\sqrt{\zeta}. \tag{3.2}$$

---

[1] Alternatively, one can assume this to be true and verify it at the end.

Then, the image of V in the s-plane is the domain $\text{Im } s < 0$. Also, the map $w \to \frac{1}{2}((w/\sigma) + (\sigma/w))$ takes the hodograph onto the lower half-plane. We now invoke the hypothesis that the map from V to H is one-to-one. It is known[1] that the most general one-to-one conformal mapping from the lower half-plane to itself has the form

$$s = \frac{At + B}{Ct + D} \tag{3.3}$$

where A, B, C, and D are real constants with $AD - BC > 0$. Consequently, we must have the relation (3.3) between the variable s defined by (3.2) and the variable

$$t = \frac{1}{2}\left(\frac{w}{\sigma} + \frac{\sigma}{w}\right). \tag{3.4}$$

The constants A, B, C, and D can be determined as follows. $\zeta = 0$ corresponds to $s = 0$, $w = 0$ and $t = \infty$. Therefore, $A = 0$. Also, $\zeta = \infty$ corresponds to $s = \infty$ and, by symmetry, to $w = i\sigma$, $t = 0$. Consequently, $D = 0$, and we can write

$$s = -\frac{c}{t},$$

where c is a positive constant. In terms of the variables ζ and w, this means that

$$\sqrt{\zeta} = \frac{2c}{(w/\sigma) + (\sigma/w)}.$$

Next, we use that fact that $w = d\zeta/dz$. This means that

$$dz = \frac{d\zeta}{w} = 8c^2\sigma^2 \frac{\sigma^2 - w^2}{(\sigma^2 + w^2)^3} dw.$$

Integrating, we find immediately the result that

$$z = \frac{2c^2}{\sigma}\left[\frac{\sigma w (3\sigma^2 + w^2)}{(\sigma^2 + w^2)^2} + \arctan \frac{w}{\sigma}\right], \tag{3.5}$$

where the constant of integration has been evaluated by using the fact that the points $z = 0$ and $w = 0$ correspond to each other.

The formula (3.5) is what we have been looking for. Except for the unknown constant c, (3.5) determines z as a function of w and, therefore, implicitly determines the velocity potential w as a function of z. Even c can be determined. If the plate has length l, then the points $z = \pm l/2$ lie on S.

---

[1] See Nehari, *op. cit.*

Therefore $|w| = \sigma$ there. But these points also lie on the plate, so that w must be real there. Setting $w = \sigma$, we find

$$z = \frac{2c^2}{\sigma}\left(1 + \frac{\pi}{4}\right),$$

so that we must have

$$c = \sqrt{\frac{l\sigma}{4 + \pi}}. \tag{3.6}$$

The flow we have just determined does not suffer from d'Alembert's paradox. To see this, we compute the force on the plate due to the fluid. For this purpose, we may assume the potential describing the external forces to be zero, for the external force on the plate simply is added on to the force exerted by the fluid. When $\Omega = 0$, the pressure at any point of the fluid is given by

$$\frac{1}{\varrho}p = \frac{1}{\varrho}p_0 - \frac{1}{2}|w|^2$$

where p_0 is a constant. To determine p_0, we assume that the pressure in the wake is zero. (This entails no loss in generality, since the addition of a constant to the pressure in the wake would result in addition of the same constant to the pressure in the fluid. The net force on the plate due to this change in the gage pressure would be zero.) On the boundary S, then, we must have $p = 0$. But $|w| = \sigma$ on S, so that

$$p = \tfrac{1}{2}(\sigma^2 - |w|^2) \tag{3.7}$$

at any point of the fluid.

The force on the plate can now be computed from (2.2), (3.5), (3.6), and (3.7). Since w is real on the plate, the result is that the component of the force parallel to the plate is zero, while the component normal to the plate is

$$-\int_{-l/2}^{l/2} p\,dz = -2\int_0^l \left[\frac{1}{2}(\sigma^2 - w^2)\right]\left[8c^2\sigma^2\,\frac{\sigma^2 - w^2}{(\sigma^2 + w^2)^3}\right]dw$$

$$= -\pi\sigma c^2 = -\frac{4l\sigma^2}{4 + \pi}. \tag{3.8}$$

This force orthogonal to the plate is called the *drag*.

Exercises

3.1   Prove that the complex velocity potential derived in § 3 is unique.

3.2   Verify that the various hypotheses made in the course of the derivation of the potential $f(z)$ in § 3 are satisfied.

3.3   Find the wake flow and compute the forces on a flat plate at an angle α to the velocity at infinity[1].

4   Discussion of the result

With the simple idea of allowing for a wake, d'Alembert's paradox has been disposed of, but there remains the question of how well the results of § 3 agree with experiment. Without further ado, we can give the answer: the force formula (3.8) gives a value for the drag at low speeds less than one-half of the measured values. The reason for the error is a considerable wake underpressure that occurs in actual flows.

We discuss this problem further in a moment, but first I would like to describe certain more agreeable features of the "wake flow" we have just found. In addition to the fact that it suffers from the d'Alembert paradox, there are two[2] things wrong with the earlier potential flow past a plate that we found in chapter five. The first is that, contrary to our ordinary experience that one can shelter from a wind in the lee of a wall, the flow of Figure 10c is symmetrical. It makes no difference on which side of the wall you may stand, the speed of the wind is the same. This qualitative paradox has been resolved by our consideration of wake flow. In the flow of Figure 16, the wall provides shelter.

The other paradox of potential flows without wakes is associated with what happens near the edges of the plate. The complex velocity potential of a flow broadside to a plate and without a wake can be seen from the results of chapter five to be

$$f(z) = \sqrt{1 + z^2},$$

if the ends of the plate are at the points $z = \pm i$. Therefore,

$$f'(z) = \frac{z}{\sqrt{1 + z^2}}$$

and this blows up at $z = \pm i$. Therefore, the speed of the flow approaches infinity near the ends of the plate, and this must mean that the hypotheses that we have made are violated.

---

[1] See Garrett Birkhoff, *op. cit.*

[2] At least. For a detailed discussion of the various paradoxes which occur in potential flows, see Garrett Birkhoff, *op. cit.*

For wake flow, on the other hand, the ends of the plate are also the ends of the free streamlines S. Consequently, the speed of the flow at the ends of this plate must be the same as the speed at infinity, and this is finite.

These agreeable features notwithstanding, there remains the problem that the computed force on a plate is far too small. As I said earlier, this is associated with an underpressure in the wake that has the effect of drawing together the two branches of the streamline S. This may have the effect of making the cavity finite instead of infinite. For a discussion of these matters, see the book by G. Birkhoff and E. Zarantonello[1].

5 Water waves

The present section does not belong in a chapter called "d'Alembert's paradox". However, I interject it here because we have been talking about flows with "free" boundaries (the streamline S), and because the entire subject of water waves is an excellent example of what I referred to earlier as the "primitive" state of continuum mechanics.

We begin by deriving the equations satisfied by an ideal fluid under gravity. These equations form the model of water waves that is usually considered in the subject. The equations are derived for two-dimensional motions; the generalization to three dimensions is obvious.

In the problem of water waves, we consider a flow between two curves $y = \eta\,(x, t)$ and $y = -h(x)$, so that the domain of the water is described by the pair of inequalities

$$-h(x) < y < \eta\,(x, t). \tag{5.1}$$

We assume that the flow is derived from a potential, so that there is a function $\varphi\,(x, y, t)$ defined in the domain (5.1) and satisfying there the equation

$$\nabla^2\varphi = \varphi_{xx} + \varphi_{yy} = 0. \tag{5.2}$$

The curve $y = -h(x)$ is thought of as the bottom. There should be no flow through it, which requires that the normal derivative be zero there:

$$\varphi_y = -h_x\varphi_x \quad \text{when} \quad y = -h(x). \tag{5.3}$$

The curve $y = \eta\,(x, t)$ is the free surface of the water. As such, it is not known *a priori*, and must be determined as part of the problem. Two conditions are, therefore, imposed on this free surface. The first is that it is composed of particles moving with the fluid. This means that if $(x(t), y(t))$ is a point on the free surface, so that

$$y(t) = \eta\,(x(t), t),$$

---

[1] Jets, wakes, and cavities. Academic Press, 1957, New York.

then differentiating with respect to t, we must have

$$\varphi_y = \eta_t + \eta_x \varphi_x \quad \text{when} \quad y = \eta. \tag{5.4}$$

This is one of the boundary conditions to be satisfied at the free surface.

The other is that the pressure at the free surface be constant, equal to atmospheric pressure. Now, in the presence of a uniform gravitational field, Bernoulli's equation is

$$gy + \frac{1}{\varrho} p + \varphi_t + \frac{1}{2} |\nabla\varphi|^2 = 0.$$

At the free surface, when $y = \eta$, we want p to be constant. Therefore, we obtain for the other boundary condition on φ the equation

$$g\eta + \varphi_t + \tfrac{1}{2}|\nabla\varphi|^2 = \text{constant} \quad \text{when} \quad y = \eta. \tag{5.5}$$

The boundary value problem (5.2–5) is extremely difficult, mostly because (5.5) is a *nonlinear* condition, and one that is imposed at an *unknown* boundary. Some idea of the difficulty of the problem may be obtained by asking what is known about it. The simplest nontrivial statement that a mathematician can make about a physical problem is that it has a solution. There are only *five* situations in which this statement can be made about the problem (5.2–5). They are as follows.

(1) $h = \infty$. In 1925, Levi–Civita[1] proved that in water of infinite depth, there is a periodic wave that progresses without change of shape. This last means that φ does not depend on x and t separately, but only on a combination $x - ct$ for some constant c. η also depends only on $x - ct$, while φ and η are both periodic functions of $x - ct$.

(2) $h = \text{constant}$. Shortly after Levi–Civita proved his result, Struik[3] showed that it could be generalized to the case of a flat horizontal bottom. Again, Struik proved the existence of a periodic wave progressing without change of shape.

(3) The solitary wave. There is a long gap between Struik's result and the next. In 1954, Friedrichs and Hyers[2] proved, again when h is constant, the existence of another type of wave, again progressing without change of shape at constant speed. This is called the *solitary wave* and can be looked on as a periodic wave à la Struik, but with infinite wavelength.

---

[1] T. Levi-Civita, Détermination rigoureuse des ondes permanents d'ampleur finie. Math. Annalen, v. 93 (1925) pp. 264–314.

[2] D. J. Struik, Détermination rigoureuse des ondes irrotationelles periodiques dans un canal à profondeur finie. Math. Annalen, v. 95 (1926) pp. 595–634.

[3] K. O. Friedrichs and D. H. Hyers, The existence of solitary waves. Communications on Pure and Applied Mathematics, v. 7 (1954) pp. 517–550.

(4)   Waves over a periodic bottom. If the bottom is periodic and has only one maximum and one minimum per period, Gerber[1] proved that there is a *steady* flow in which the free surface has the same properties. In addition, the troughs of the free surface lie directly over the troughs of the bottom, and the crests lie over the crests.

(5)   Flows over a monotone bottom. In the same paper[1] Gerber proved that over a monotone bottom, there is a flow with a monotone free surface. Again this can be looked upon as a flow over a periodic bottom with infinite period.

It should be noted that all these examples are essentially examples of *steady* flows. The last two are steady to begin with. The first three become steady when observed from a coordinate system moving with the speed c. There are *no* known non-steady flows.

All the above flows are two-dimensional. There are *no* known three-dimensional flows.

All the above flows are for very special bottom shapes. There are *no* known theorems about existence of flows over "general" bottoms.

For a fuller discussion of water waves, see the book on the subject by J. J. Stoker[2].

---

[1] Robert Gerber, Sur les solutions exactes des équations du mouvement avec surface libre d'un liquide pesant. Journal de Mathématiques. 34 (1955) 185–299.

[2] Water waves. Interscience publishers, New York 1957.

CHAPTER 6

Flows with circulation

1 The stream function

UNTIL NOW we have considered only potential flows. A consequence has been the necessity for the introduction of wakes to resolve the d'Alembert paradox. Another alternative, and one that seems more suitable for streamlined obstacles, as opposed to the plate of chapter six, is to consider flows in which the circulation is no longer zero.

If the circulation is not zero, the idea of a single-valued velocity potential has to be abandoned, and one has to consider the more complicated equations of motion (2.2.2–3). For two-dimensional flows, which is what we continue to consider, these equations, written out in full, are

$$
\left.
\begin{aligned}
u_t + uu_x + vu_y + \left(\frac{1}{\varrho}p - \Omega\right)_x &= 0, \\
v_t + uv_x + vv_y + \left(\frac{1}{\varrho}p - \Omega\right)_y &= 0, \\
u_x + v_y &= 0.
\end{aligned}
\right\}
\tag{1.1}
$$

For potential flows, we introduced the stream function as the imaginary part of the complex velocity potential. The basic property of the stream function is that the velocity at any point can be computed from the formulas

$$
\left.
\begin{aligned}
u &= \psi_y, \\
v &= -\psi_x.
\end{aligned}
\right\}
\tag{1.2}
$$

This is really a contrary way of defining the stream function, however, since in some ways it is more basic than the velocity potential. One reason is that it exists whether or not the flow is potential flow. This fact is an immediate consequence of (1.1c), which is the necessary and sufficient condition for u and v to be expressible in the form (1.2) for some function ψ.

When u and v are given by (1.2) equation (1.1c) is satisfied automatically. The other two of equations (1.1) can be written in the form

$$
\begin{aligned}
\frac{\partial}{\partial x}\left[\frac{1}{\varrho}p - \Omega + \frac{1}{2}(\psi_x^2 + \psi_y^2)\right] &= \psi_x \nabla^2 \psi - \psi_{yt}, \\
\frac{\partial}{\partial y}\left[\frac{1}{\varrho}p - \Omega + \frac{1}{2}(\psi_x^2 + \psi_y^2)\right] &= \psi_y \nabla^2 \psi + \psi_{xt}.
\end{aligned}
\tag{1.3}
$$

Differentiating the first of these equations with respect to y and the second with respect to x and subtracting the results leads to the following equation for ψ alone:

$$\left(\frac{\partial}{\partial t} + u\frac{\partial}{\partial x} + v\frac{\partial}{\partial y}\right)(\nabla^2\psi) = 0. \tag{1.4}$$

Conservely, given any solution of (1.4), where u and v are given by (1.2), we can recover the pressure by using the formula

$$p - \Omega + \tfrac{1}{2}(\psi_x^2 + \psi_y^2) = \int_{(x_0, y_0)}^{(x, y)} [(\psi_x\nabla^2\psi - \psi_{yt})\,dx + (\psi_y\nabla^2\psi + \psi_{xt})\,dy]. \tag{1.5}$$

Here, (x_0, y_0) is any fixed point in the flow; the integral is independent of path because of (1.4).

2 Circulation

The circulation Γ defined in § 2.3 depends in a very simple way on the stream function ψ. By definition of Γ, we have

$$\Gamma = \int_\gamma (u\,dx + v\,dy) = \int_\gamma (\psi_y\,dx - \psi_x\,dy), \tag{2.1}$$

where γ is any curve in the fluid. Green's theorem can be applied immediately to (2.1) to obtain the result

$$\Gamma = \int\int_S \nabla^2\psi\,dx\,dy, \tag{2.2}$$

if γ is immersed in the fluid and S is the domain bounded by γ.

Now, one class of solutions of the equation (1.4) for ψ is always available: it is the solutions of Laplace's equation

$$\nabla^2\psi = 0. \tag{2.3}$$

In a simply connected domain, (2.3) implies, via (2.2), that Γ is zero and, just as in chapter three, that there is a single-valued velocity potential. However, we shall see that for flows past bounded obstacles (in which the domain of the fluid is not simply connected) another class of potential flows, more general than any we have so far considered, can be constructed.

Let ψ be any solution of (2.3). ψ is then automatically a solution of (1.4). Moreover, (2.3) is *linear* and, therefore solutions of (2.3) can be superposed. We shall make heavy use of these facts.

Consider what was called in chapter five the source solution of (2.3), for which[1]

$$\psi = \frac{\varkappa}{2\pi} \log r.$$ (2.4)

In any domain excluding the origin, this is a solution of (2.3) and, therefore, of (1.4). We compute the circulation associated with (2.4). If γ is any simple closed curve that does not encircle the origin, we can use (2.2) to conclude that the associated circulation is zero.

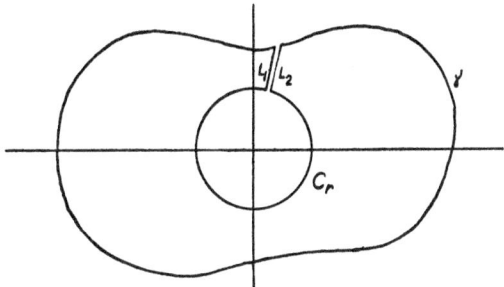

Figure 18

Now, let γ be any simple closed curwe that does encircle the origin. Let C_r be the circle of radius r centered at the origin. If the curves γ and C_r are connected by two neighboring line segments L_1 and L_2, the curve $\gamma \cup L_1 \cup C_r \cup L_2$ does not encircle the origin. Therefore, the circulation around this curve is zero. Since the integrals along L_1 and L_2 cancel as these lines approach each other, we see that the circulation around γ is the same as that around C_r.

But the circulation around C_r is

$$-\int_0^{2\pi} \psi_r r \, d\theta = -\varkappa.$$

Thus, the stream function (2.4) has the property that the associated circulation is zero for any curve that does not encircle the origin and is the constant $(-\varkappa)$ for any curve that does.

The flow corresponding to the stream function (2.4) is simple to understand. Because of (1.2) the radial velocity at any point in the flow is

$$\frac{1}{r} \psi_\theta = 0,$$ (2.5)

---

[1] We write $-\varkappa$ instead of μ for later convenience in signs.

while the angular velocity is

$$-\psi_r = -\frac{\varkappa}{2\pi r}. \tag{2.6}$$

Thus, (2.4) is the stream function of a circulatory flow having no radial velocity at any point. We have seen this flow before when we looked at Rankine's combined vortex (§ 2.4(4)).

(2.4) is a solution of Laplace's equation. Let us add it to the stream function associated with the potential flow around a cylinder. (§ 5.2(4)). If the cylinder has radius 1, the result is the function

$$\psi = U\left(r - \frac{1}{r}\right)\sin\theta + \frac{\varkappa}{2\pi}\log r. \tag{2.7}$$

The first part of this function is the stream function found in § 5.2(4); the second part is just (2.4).

It is not hard to visualize the flow associated with (2.5). Notice first that for any value of $\varkappa$, ψ is constant when $r = 1$. Therefore, the normal velocity to the curve $r = 1$ is zero, and we may replace this curve by a fixed wall. Thus, for any $\varkappa$, (2.5) is the stream function of a flow past the cylinder $r = 1$. Moreover, the velocity at infinity associated with the circulatory part, $- (\varkappa/2\pi)\log r$, of ψ is zero. This is seen by a glance at (2.5) and (2.6). Thus, for all values of $\varkappa$, the velocity at infinity of the flow with stream function (2.7) is $(U, 0)$.

The details of the flow become clear enough that it can be sketched if one considers what happens to the stagnation points as a function of $\varkappa$. At such points, the velocity is zero. When $\varkappa = 0$, the points S_1 and S_2 of Figure 19 are stagnation points.

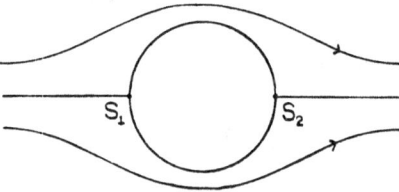

Figure 19

A point is a stagnation point if and only if both ψ_r and ψ_θ are zero According to (2.7), then, this means that

$$\left.\begin{aligned} U\left(1 + \frac{1}{r^2}\right)\sin\theta &= -\frac{\varkappa}{2\pi r}, \\[2mm] U\left(r - \frac{1}{r}\right)\cos\theta &= 0. \end{aligned}\right\} \tag{2.8}$$

The second of these equations is always satisfied when $r = 1$. In that case there will be a stagnation point when

$$\sin \theta = -\frac{\varkappa}{4\pi U}. \tag{2.9}$$

For $0 < \varkappa < 4\pi U$ there are two such values of θ, and if $\varkappa$ is small, one is negative and near zero, and the other is near $-\pi$. Thus for small values of $\varkappa$, both S_1 and S_2 move down to produce the flow sketched in Figure 20. The important thing to notice about Figure 20 is that it lacks one axis of symmetry that the flow sketched in Figure 19 possessed. We return to this point in a moment.

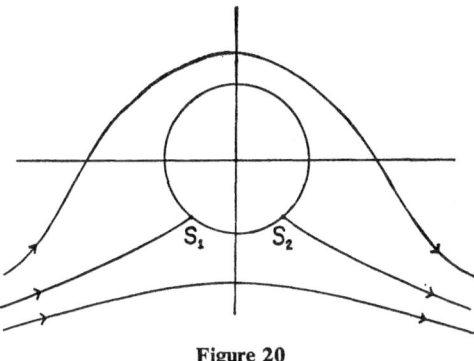

Figure 20

When $\varkappa > 4\pi U$, (2.9) becomes impossible. But we notice that (2.8b) is still satisfied when $\theta = -\pi/2$, whether or not $r = 1$. When $\theta = -\pi/2$, however, (2.8a) becomes

$$1 + \frac{1}{r^2} = \frac{\varkappa}{2\pi U r}.$$

This is a quadratic equation in $1/r$, and it has two solutions. Since $\varkappa > 4\pi U$, both are real. One of them is such that $r < 1$, and so lies outside the flow. The other lies in the flow. What is happening here is this. When $\varkappa$ is large, the circulatory part of the flow dominates near the cylinder. Thus, part of the fluid simply swirls around the cylinder. This part is separated from the rest of the flow by a self-intersecting streamline. At the intersection point, the speed is necessarily zero, and this is the stagnation point. The flow is sketched in Figure 21.

The asymmetry of the flows sketched in Figures 20 and 21 has a crucial consequence. The cause of the asymmetry is clear physically. We have added a circulatory (clockwise) flow to the potential flow obtained by setting $\varkappa = 0$. This results in a net *increase* in speed above the cylinder, and a net *decrease* below it. Because of Bernoulli's equation, then, the pressure has been *decreased* above and *increased* below the cylinder from the values it had when $\varkappa$ is zero. Since there are no forces on the cylinder when $\varkappa$ is zero, we conclude that the flow associated with the stream function (2.7) must exert a net *lift* on the cylinder. Our next subject of study will be this force.

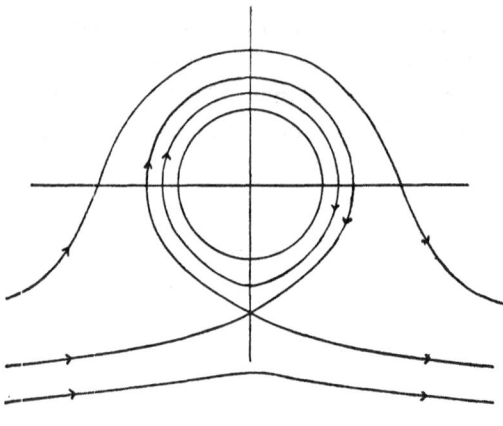

Figure 21

3 Circulatory flow past an airfoil

But first, we have to see how to create a flow past an arbitrary bounded object similar to the one just introduced for a cylinder.

Because the circulation associated with the stream function (2.7) and any curve that does not encircle the cylinder is zero, the same argument that we used in proving Lagrange's theorem shows that in any simply connected domain consisting entirely of fluid particles, there is a velocity potential. The difference between the potential here and the ones we have discussed earlier is that this one is no longer single-valued[1]. When one takes a turn about the cylinder, the velocity potential changes its value.

We require the complex velocity potential associated with the stream function (2.7). This means that we require an analytic function having (2.7) for its imaginary part. Inspection shows that the following function will do:

$$f(z) = U\left(z + \frac{1}{z}\right) - \frac{\varkappa}{2\pi i}\log z. \tag{3.1}$$

---

[1] As we shall see, this does not violate the tacit hypothesis we have been making all along that there is a single-valued velocity vector associated with each point in the fluid, since the derivatives of the velocity potential, unlike the potential itself, are single-valued. Therefore, associated with each point in the fluid there is still a single-valued velocity.

To create a circulatory flow past a body, we use the same idea that was used in proving theorem 5.3.1. Let V be the exterior of the body, and let $k(z)$ map V onto the exterior of the unit disk. In addition, let $k(z)$ have a simple pole at infinity such that

$$k(\infty) = \infty, \quad k'(\infty) > 0. \tag{3.2}$$

The assertion is then that if f is given by (3.1) the function

$$F(z) = f(k(z))$$

is a weak complex velocity potential for the domain V.

In fact, when z is in V, $|k(z)| > 1$, so that F is analytic. When z approaches the boundary of V, $k(z)$ approaches the boundary of the unit disk, and so Im $F(z)$ goes to a constant. Finally,

$$F'(z) = f'(k(z)) \, k'(z)$$

and, as $|z| \to \infty$,

$$F'(z) \to Uk'(\infty),$$

as (3.1) and (3.2) show.

This proves the following analog of theorem 5.3.3.

THEOREM 3.1   *Let V be an exterior domain with a connected boundary and at least two boundary points. Let $k(z)$ be an analytic function in V, having a simple pole at infinity, mapping V conformally onto the exterior of the unit disk and having the properties*

$$k(\infty) = \infty, \quad k'(\infty) > 0.$$

Then, the function

$$U\,[k(z) + 1/k(z)] - \frac{\varkappa}{2\pi i} \log k(z)$$

is a weak complex velocity potential of a flow in V. The velocity at infinity of this flow is horizontal and has magnitude $Uk'(\infty)$.

Naturally, there is nothing supernatural about the horizontal direction. Let the desired velocity at infinity be the vector (U, V). Write $S = U + iV$ and let $\bar{S}$ be the complex conjugate of S. A review of the argument leading to theorem 3.1 then shows that a weak complex velocity potential for the domain V of the theorem is

$$\frac{\bar{S}}{k'(\infty)} \, [k(z) + 1/k(z)] - \frac{\varkappa}{2\pi i} \log k(z). \tag{3.3}$$

Incidentally we remark that this formula shows, as predicted in the footnote on p. 94, that there is a single-valued velocity associated with each point of V, since the derivative of (3.3) is single-valued.

4   Lift. Blasius' theorem

When $\varkappa \neq 0$, the flows described in the theorem just proved are not subject to d'Alembert's paradox. Of course, this is to be expected because of the discussion in § 2. In this section we evaluate the forces on an airfoil quantitatively.

Let V be an exterior domain with a connected, smooth boundary. Let $f(z)$ be a complex velocity potential in V. Denote the complement of V by B; we think of B as a body immersed in the fluid. The force exerted by the fluid on B is given by (6.2.2):

$$F = -\int_{\partial V} pn \, d\lambda,$$

where λ is arclength on ∂V and n is the normal to V directed out of the fluid.

For the purposes of this section, it is convenient to represent all vectors as complex numbers. Thus, if we have a vector (α, β), we write $(\alpha, \beta) = \alpha + i\beta$. A bar over a quantity denotes its complex conjugate. In this notation, the conjugate of the force on B is

$$\bar{F} = -\int_{\partial V} p\bar{n} \, d\lambda. \tag{4.1}$$

Equation (1.5) (which is just Bernoulli's equation in the present context) says that if the external forces are zero,

$$\frac{1}{\varrho} p = \frac{1}{\varrho} p_0 - \frac{1}{2} |f'|^2,$$

where p_0 is a constant. Clearly the contribution of p_0 to the force is zero. Therefore, we have

$$F = \frac{\varrho}{2} \int_{\partial V} |f'|^2 \, \bar{n} \, d\lambda.$$

Since the velocity normal to ∂V is zero, the velocity itself is tangent to ∂V. Thus, if θ denotes the angle between the velocity vector and the real axis, we have

$$n = e^{i(\theta - \pi/2)}$$

(see Figure 22). Thus, we have

$$\bar{F} = \frac{i\varrho}{2} \int_{\partial V} |f'|^2 \, e^{-i\theta} \, d\lambda.$$

Finally f', being the derivative of the complex velocity potential, is the conjugate of the velocity vector. Therefore,

$$f' = |f'| \, e^{-i\theta}.$$

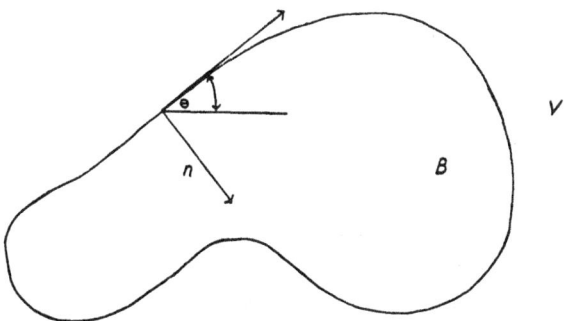

Figure 22

This gives

$$F = \frac{i\varrho}{2} \int_{\partial V} f'^2(z) \, e^{i\theta} \, d\lambda = \frac{i\varrho}{2} \int_{\partial V} f'^2(z) \, dz. \qquad (4.2)$$

In addition to a force the fluid exerts a moment tending to rotate B, given by

$$-\frac{\varrho}{2} \operatorname{Re} \int_{\partial V} z f'^2(z) \, dz; \qquad (4.3)$$

this is proved in the same way (4.1) was.

Now, $f(z)$ has the form (3.3). Also $k(z)$ has a simple pole at infinity, so that

$$k(z) = k'(\infty) \, [z + l(z)],$$

where $l(z)$ is analytic at infinity. Since the derivative of a function analytic at infinity is $0 \, (1/z^2)$ there, this shows that

$$f'(z) = \bar{S} - \frac{\varkappa}{2\pi i z} + O\left(\frac{1}{z^2}\right) \qquad (4.4)$$

in a neighborhood of infinity. The function $f'^2(z)$ is analytic in V. Therefore, by Cauchy's theorem, the contour ∂V can be distorted to a large circle without changing the value of (4.2). Evaluating (4.2) along a large circle and letting the radius of the circle go to infinity, we find, using (4.4),

$$F = i\varrho\bar{\varkappa}S. \qquad (4.5)$$

A similar result can be derived[1] for the moment (4.3).

THEOREM 4.1 (Blasius)[2] *Let V be an exterior domain with a connected, smooth boundary. A potential flow in V with circulation $(-\varkappa)$ and velocity S at infinitiy*

---

[1] See Horace Lamb, Hydrodynamics. Dover Publications, New York, 1945.

[2] H. Blasius, Funktiontheoretische Methoden in der Hydrodynamik. Zeit. für Math. und Physik, 58 (1910) 90–110.

exerts a force on the complement of V given by

$$F = i\varrho \bar{\varkappa} S.$$

A consequence of Blasius' theorem about which a great deal of nonsense has been written is

COROLLARY 4.2 *Let V be a domain as the theorem 4.1, and let B denote its complement. Then the force exerted by the fluid on B is independent of the shape of B and its orientation with respect to the direction of the flow at infinity. If the circularion* ($-\varkappa$) *is real, the force is orthogonal to this direction.*

That the force is independent of B and its orientation is clear, for as (4.5) shows, the force depends only on ϱ, $\varkappa$, and S. The last sentence is also clear, for if ϱ and $\varkappa$ are real, the vectors S and $i\varrho\varkappa S$ are easily seen to be orthogonal.

The hypotheses in Blasius' theorem and its corollary must be respected. Consider, as in § 5.4(5), a flow past a flat plate making an angle α with the negative real axis. If the flow velocity at infinity is horizontal and the circulation is real, corollary 4.2 seems to say that the force due to the flow past the three plates shown in Figure 23 is the same for all three plates and, what is more, that it is directed straight up!

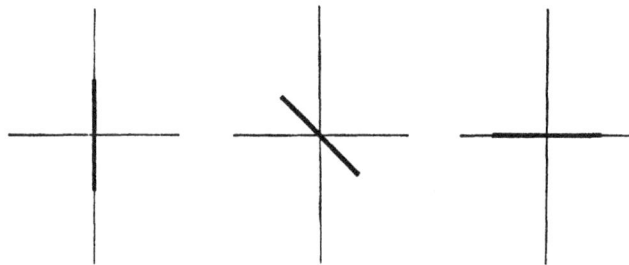

Figure 23

Of course if one computes the force on any of the plates of Figure 23 directly by means of (4.1) he finds that the force due to a circulatory flow about any of them is orthogonal to the plate, as it must be for any ideal fluid flow[5]. The apparent contradiction with the conclusion of corollary 4.2 is due to the fact that the domains V of Figure 23 do not have smooth boundaries as required by Blasius' theorem.

Unfortunately this means, once again, that Blasius' theorem is *not* a physically meaningful result. For, if one considers a flow past a very thin body, nearly a plate, but one which has a continuously turning normal at each point, Blasius' theorem says that the force on the body is orthogonal to the

---

[5] This is a consequence of the hypothesis that the reduced stress tensor is zero.

velocity at infinity, while in the limit as the thickness of the body goes to zero, the force is orthogonal to the resulting plate. This shows that the force on the body is not a continuous function of its thickness, which means that a very small change in the shape of a body can result in a large change in the force on it. Such a conclusion can never be physically significant.

Exercise

4.1 Evaluate the force in circulatory flow on a flat plate making an arbitrary angle with the direction of the flow at infinity using (4.1). Show, in particular, that the force is orthogonal to the plate.

5 Joukowski's hypothesis. Goodbye to all that

The idea of circulatory flow was introduced into fluid mechanics in the years around 1910. The date is significant because it was the time of the advent of the airplane and the computation of forces on wings was important for design of such machines. Cavity flow, while explaining drag on a body, says nothing at all about lift. In flows with circulation, however, forces orthogonal to the flow at infinity (*lift* forces) appear very naturally.

But so does a new paradox. We created in theorem 3.1 a one-parameter family of flows past a given body, the parameter being the circulation. But when a body is immersed in a flow, it does not request that the flow produce the particular value of the circulation that would tickle its fancy that day. Thus, in contrast to the situation in d'Alembert's time when there were no flows known that exerted forces on bodies, by 1910 there were too many such flows. How is the circulation to be chosen? In 1910, Joukowski offered a brilliant solution to this problem, at least for bodies shaped like the usual airfoils.

Such a body has a shape roughly like that of Figure 14, smooth except for a single point at the trailing edge. (The point z_0 in Figure 14.) Our experience so far leads us to believe that there exist flows past such bodies with continuous derivatives in the closure of the domain of the fluid, except possibly for the point z_0 on the trailing edge where ∂V is not smooth. At z_0, the angle (measured through the fluid) between the upper and lower surfaces of the wing exceeds π. Therefore, theorem 4.2.1 leads us to believe that the velocity of the flow is generally infinite at z_0. Joukowski suggested that the value of the circulation be adjusted in such a way that the velocity at the trailing edge be bounded.

Generally, this adjustment can be achieved. According to theorem 3.1, a circulatory complex velocity potential past a wing is given by the formula (3.3). The speed, then, is the absolute value of the derivative of (3.3), and this derivative is

$$\left\{ \frac{\bar{S}}{k'(\infty)} \left[1 - \frac{1}{k^2(z)} \right] - \frac{\varkappa}{2\pi i k\,(z)} \right\} k'(z). \qquad (5.1)$$

Let z_0 be the point on the trailing edge of the wing. Since $k(z)$ maps the domain of the fluid onto the domain $|w| > 1, |k(z_0)| = 1$. Therefore, $k(z_0) = e^{i\beta}$, where $\beta = \arg k(z_0)$ is real. Set

$$\varkappa = -\frac{4\pi\bar{S}}{k'(\infty)} \sin\beta; \qquad (5.2)$$

then the quantity multiplying $k'(z)$ in (5.1) vanishes when $z = z_0$. Thus, the choice (5.2) for $\varkappa$ will produce the desired effect of bounding (5.1) if the singularity of $k'(z)$ at z_0 is not too severe.

Notice: the Joukowski hypothesis suggests an *ad hoc* value for the circulation valid only for wings having a single sharp trailing edge. There is no theoretical justification for it, and there is no generalization to be applied if the trailing edge is ever so slightly rounded. *Yet Joukowski's theory is the most well known and widely used theory of lift extant.* There is work to be done!

How does a wing in a fluid select that unique value of the circulation that makes the velocity everywhere finite (if that is indeed what happens)? If that *is* what happens, what value does the circulation have if the wing has no sharp trailing edge? The circulatory flows we have studied are steady, which means that the equations of motion possess steady solutions, but do steady flows ever occur in nature, or have we been pursuing a fantasy all along? If steady flows do occur, which ones occur? Are they stable, or will a small perturbation of the flow cause it to drift to another steady solution, or even an unsteady one? The answer to none of these questions is known.

All the answers undoubtedly involve a study of the *initial value* problem (3.7.1–4). When a body is inserted in a flow, it is too much to expect the flow to be steady to begin with. The best one can hope for is that the flow be asymptotic to a steady one for large time. The initial value problem has a unique solution, as we saw in § 3.7 and exercise 3.7.1. Presumably what happens when a wing with a sharp trailing edge is inserted in a flow, is that there ensues a period of unsteady flow, following which the circulation converges to the unique value (5.2). For bodies without sharp trailing edges, the circulation converges to whatever value it converges to, and presumably this can be computed by examining carefully what happens to the solution of the initial value problem as $t \to \infty$.

Reasonable enough, isn't it? The argument has been made before, but it's all impossible. Consider the following experiment. Insert a wing in a flow initially at rest. Then, after the position of the wing has been established, begin to blow air over it, the velocity at infinity initially being zero and smoothly increasing to a limiting value. The circulation about any curve in the fluid is initially zero. According to lemma 2.3.4, then, it must stay that way, and no circulation can ever develop. Which only shows that if the questions we asked are to be answered at all and, in particular, if the Joukowski hypothesis is to be proved correct in some asymptotic sense, the hypothesis of ideal fluid must be abandoned.

This does not mean that the work of the past five chapters is worthless. We began in chapter one with a system of particles and then defined a fluid as a system satisfying the Stokes hypothesis. No heuristic justification was given, and it was necessary to show that the hypothesis, even in the very simple form it takes for an ideal fluid, describes phenomena that are in some degree in accord with our intuition about how fluids behave. I think that has been achieved. The flow of Figure 2, for example, looks more or less what we expect the flow past a sphere to look like. However, the d'Alembert paradox, for instance, shows that ideal flows are not perfect models of real ones. No models ever are, but having come this far, it is important to refine the model at least enough that it becomes possible to provide an explanation of the d'Alembert paradox that is not obviously inconsistent, as was the explanation provided two paragraphs back. Such a refinement is what we turn to now.

Exercise

5.1   Consider a wing such as the one in Figure 14 which is smooth except for the single point z_0. Suppose the angle α between the top and the bottom surface (measured through the fluid) satisfies $\pi/2 < \alpha < \pi$.

(a)   Show that if the circulation is chosen to satisfy (5.2) then $|f'(z)|$ is bounded in the closure of V.

(b)   Prove that the force on the wing can be computed from Blasius' formula (4.5).

CHAPTER 7

Viscous fluids

1 The Stokes hypothesis again

THE DIFFICULTY with ideal fluids, and the source of the d'Alembert paradox, is that in such fluids there are no frictional forces. Two neighboring portions of an ideal fluid can move at different velocities without rubbing on each other, provided they are separated by a streamline. For example, we may think of a fluid filling the entire plane. In the upper half-plane, the flow can be uniform, and in the lower half-plane, the velocity can be everywhere zero. If the flow is ideal, the fluid in the upper half-plane will never pull the fluid in the lower along with it. It is precisely this fact that allowed us to replace a portion of the plane bounded by a streamline by a rigid wall in earlier chapters.

It is clear that such a phenomenon can never occur in a real fluid, and the question is how frictional forces can be introduced into a model of a fluid. The thing to understand is that frictional forces are caused by *relative* motion of different parts of a fluid. We wish to introduce a model in which a portion of a fluid exerts no forces on a neighboring portion if they are not moving relative to each other, but in which forces are exerted if there is relative motion. This understanding leads directly to the Stokes hypothesis.

The argument is one often used in physics and engineering and is well worth appreciating. It is based on a modification of Occam's razor: *do not complicate your hypotheses unnecessarily*. This means that if, under the simplest hypotheses consistent with the phenomena we wish to describe, a model can be derived that is already too difficult to cope with completely, then this model should be retained unless and until paradoxes appear with in it.

We wish to apply this to derive the form of the reduced stress tensor, keeping in mind that there are to be no forces between portions of a fluid that are not in relative motion. This means, first, that the reduced stress tensor σ is to be zero when there is no relative motion. To begin, then, σ must be zero when there is no motion at all—that is, when the velocity s is zero. Moreover, σ must be zero when the velocity s is constant throughout the fluid, for then there is no relative motion between different parts of the fluid. The simplest hypothesis that assures us of this is that σ is a function of the derivatives of s only:

$$\sigma = \sigma\,(\nabla s,\ \nabla\nabla s,\ \dots); \tag{1.1}$$

here, the function on the right is supposed to vanish when all its arguments are zero.

Within the framework of the hypothesis (1.1), the simplest assumption that can be made is that σ depends only on the *first* derivatives of s. Again, the simplest hypothesis about the form of this dependence is that it is *linear*. When these hypotheses are made, each component of the matrix σ has the form

$$\sigma_{ij} = a_{ij}u_x + b_{ij}u_y + c_{ij}u_z + d_{ij}v_x + e_{ij}v_y + f_{ij}v_z$$
$$+ g_{ij}w_x + h_{ij}w_y + k_{ij}w_z, \tag{1.2}$$

where coefficients of the derivatives are constants. We show now that (1.2) already implies a restricted form of the Stokes hypothesis.

The coefficients in (1.2) still have to be adjusted so that σ is zero when there is no relative motion between different portions of a fluid. Now, in a fluid rotating like a rigid body with constant angular velocity, there is no relative motion between its various parts. Consider such a fluid, with the z-axis the axis of rotation. In this case, it is not hard to show that the velocity of the fluid at the point $q = (x, y, z)$ is

$$s = \omega \left(-y, x, 0\right),$$

where ω is a constant. By (1.2), then,

$$\sigma_{ij} = \omega \left(d_{ij} - b_{ij}\right),$$

and this must be zero, so that

$$d_{ij} = b_{ij}.$$

In a similar way, making the axis of rotation the x- and then the y-axis, we find

$$g_{ij} = c_{ij},$$
$$h_{ij} = f_{ij}.$$

Therefore,

$$\sigma_{ij} = a_{ij}u_x + e_{ij}{}^l{}_y + k_{ij}w_z + b_{ij}\left(u_y + v_x\right) + c_{ij}\left(u_z + w_x\right)$$
$$+ f_{ij}\left(v_z + w_y\right). \tag{1.3}$$

This is the Stokes hypothesis, for the quantities $u_x, v_y, w_z, u_y + v_x, u_z + x_x,$ $v_z + w_y$ occurring in (1.3) are just the entries in the deformation tensor (1.4.23), so that σ is a function of the deformation tensor D. Of course, (1.3) is linear in the components of D, which is more than is required by the Stokes hypothesis. The linearity of (1.3) is a consequence of our diligent application of Occam's razor: we took as the simplest hypothesis that σ is a linear function of ∇s. Some mathematicians, while accepting the other hypotheses, have been offended by the assumption of linearity, and perhaps they are correct. That is where the last part of the razor plays a role. The model should be retained *until contradictions appear*. Until they do, consistency requires that the simplest hypotheses be made. We have more to say about this in § 5.

The relation (1.3) can be simplified. A fundamental hypothesis made in any analysis of a physical problem is that equations representing a physical phenomenon must be independent of the coordinate system used to describe it. Let C and C' be two coordinate systems, C' being obtained from C by a rotation. It is well known that any such change of coordinates is defined by an orthogonal matrix S such that any vector having the components

$$ n = \begin{pmatrix} n_1 \\ n_2 \\ n_3 \end{pmatrix} $$

in C has the components Sn in C'. If Σ is any smooth surface in a fluid, then except for a sign, the force exerted by the fluid across Σ is (cf. § 1.7)

$$ F = \int_\Sigma \tau n \, dA, $$

where n is the normal to Σ, and τ is the stress tensor. Denote the force F and the matrix τ in C' by primes. Then we have

$$ F' = \int_\Sigma \tau' n' \, dA = \int_\Sigma \tau' Sn \, dA. $$

Now, F is a vector, so that $F' = SF$. This means that

$$ F' = \int_\Sigma \tau' Sn \, dA = SF = \int_\Sigma S\tau n \, dA. $$

Thus if we begin with a coordinate system C in which the matrix τ is given, and then rotate the system, τ must be changed to a matrix τ', where τ and τ' are related by

$$ S\tau n = \tau' Sn, $$

where S is the orthogonal matrix specifying the rotation and, since Σ is arbitrary, n is any unit vector. Since the rotation is also arbitrary, this clearly entails

$$ \tau' = S\tau S^{-1} \tag{1.4} $$

for any orthogonal matrix S. Now,

$$ \tau = pI + \sigma, $$

where I is the identity matrix. Since I commutes with every matrix, (1.4) can only hold if

$$\sigma' = S\sigma S^{-1} \tag{1.5}$$

for any orthogonal matrix S.

Let a, b, ... be the matrices (a_{ij}), (b_{ij}), ... Then, (1.3) says

$$\sigma = au_x + ev_y + kw_z + b\,(u_y + v_x) + c\,(u_z + w_x) + f\,(v_z + w_y). \tag{1.6}$$

Let the coordinates of a point be (x, y, z) in C and (x', y', z') in C'. Similarly, let the velocity s, which has the components (u, v, w) in C, have the components (u', v', w') in C'. Then we have

$$\sigma' = au'_{x'} + ev'_{y'} + kw'_{z'} + b\,(u'_{y'} + v'_{x'}) + c\,(u'_{z'} + w'_{x'})$$

$$+ f\,(v'_{z'} + w'_{y'}). \tag{1.7}$$

Let C' be obtained from C by rotation of $180°$ about the z-axis. Then,

$$x' = -x, \quad y' = -y, \quad z' = z,$$

while

$$u' = -u, \quad v' = -v, \quad w' = w$$

(see Figure 24).

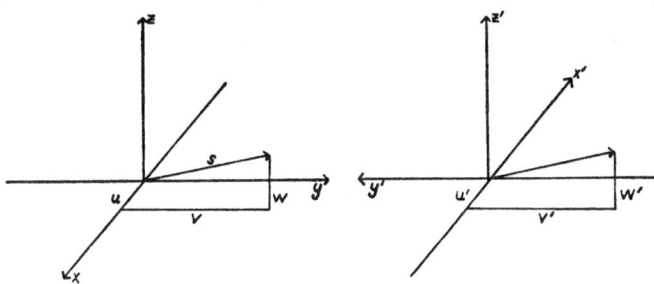

Figure 24

By (1.7), then,

$$\sigma' = au_x + ev_y + kw_z + b\,(u_y + v_x) - c\,(u_z + w_x) - f\,(v_z + w_y). \tag{1.7}$$

For this particular rotation, S has the form

$$S_1 = \begin{pmatrix} -1 & 0 & 0 \\ 0 & -1 & 0 \\ 0 & 0 & 1 \end{pmatrix},$$

and if $\lambda = (\lambda_{ij})$ is any matrix,

$$S_1 \lambda S_1^{-1} = \begin{pmatrix} \lambda_{11} & \lambda_{12} & -\lambda_{13} \\ \lambda_{21} & \lambda_{22} & -\lambda_{23} \\ -\lambda_{13} & -\lambda_{32} & \lambda_{33} \end{pmatrix}. \tag{1.8}$$

Applying this to σ as given by (1.6), and using (1.5) and (1.7), we find that

$$S_1 a S_1^{-1} = a, \qquad S_1 b S_1^{-1} = b,$$

$$S_1 e S_1^{-1} = e, \qquad S_1 c S_1^{-1} = -c,$$

$$S_1 k S_1^{-1} = k, \qquad S_1 f S_1^{-1} = -f.$$

The form of (1.8) shows that these equations can only hold if

$$a_{13} = a_{31} = a_{23} = a_{32} = 0,$$

$$e_{13} = e_{31} = e_{23} = e_{32} = 0,$$

$$k_{13} = k_{31} = k_{23} = k_{32} = 0,$$

$$b_{13} = b_{31} = b_{23} = b_{32} = 0,$$

$$c_{11} = c_{12} = c_{21} = c_{22} = c_{33} = 0,$$

$$f_{11} = f_{12} = f_{21} = f_{22} = f_{33} = 0.$$

Next, apply the same reasoning to the coordinate systems obtained from C by rotation through 180° about the y- and the x-axes. One finds that a, e, and k must be diagonal matrices while b, c, and f have the forms

$$b = \begin{pmatrix} 0 & b_{12} & 0 \\ b_{21} & 0 & 0 \\ 0 & 0 & 0 \end{pmatrix},$$

$$c = \begin{pmatrix} 0 & 0 & c_{13} \\ 0 & 0 & 0 \\ c_{31} & 0 & 0 \end{pmatrix},$$

$$f = \begin{pmatrix} 0 & 0 & 0 \\ 0 & 0 & f_{23} \\ 0 & f_{32} & 0 \end{pmatrix}.$$

All this means that σ has the form

$$\sigma = \begin{pmatrix} a_{11}u_x + e_{11}v_y + k_{11}w_z & b_{12}(u_y + v_x) & c_{13}(u_z + w_x) \\ b_{21}(u_y + v_x) & a_{22}u_x + e_{22}v_y + k_{22}w_z & f_{23}(v_z + w_y) \\ c_{31}(u_z + w_x) & f_{32}(v_z + w_y) & a_{33}u_x + e_{33}v_y + k_{33}w_z \end{pmatrix}.$$

Now repeat the reasoning again, rotating the coordinate system 90° about the z-, the y-, and then the x-axes. One finds that all the b's, c's, and f's are equal, that

$$a_{22} = a_{33} = e_{11} = e_{33} = k_{11} = k_{22}, \qquad (1.9)$$

and that

$$a_{11} = e_{22} = k_{33}. \qquad (1.10)$$

Let $(-\mu)$ be the common value of the b's, c's, and f's, α the value of (1.9) and β the value of (1.10). Then,

$$\sigma = \begin{pmatrix} \beta u_x + \alpha (v_y + w_z) & -\mu (u_y + v_x) & -\mu (u_z + w_x) \\ -\mu (u_y + v_x) & \alpha (u_x + w_z) + \beta v_y & -\mu (v_z + w_y) \\ -\mu (u_z + w_x) & -\mu (v_z + w_y) & \alpha (u_x + v_y) + \beta w_z \end{pmatrix}$$

$$= -2\mu D + \alpha (\nabla \cdot s) I + (\beta - \alpha + 2\mu) \begin{pmatrix} u_x & 0 & 0 \\ 0 & v_y & 0 \\ 0 & 0 & w_z \end{pmatrix}. \qquad (1.11)$$

The first two terms here satisfy (1.5). Therefore, the third term must. But this is impossible unless $\beta - \alpha + 2\mu = 0$. Thus, we have

$$\sigma = -2\mu D + \alpha (\nabla \cdot s) I.$$

We still wish the fluid to satisfy conservation of mass, and the condition for this is independent of the form of the stress tensor. For incompressible fluids, the condition for conservation of mass is

$$\nabla \cdot s = 0,$$

just as we had before. Therefore, finally, for incompressible fluids, the Stokes hypothesis becomes

$$\sigma = -2\mu D. \qquad (1.12)$$

The constant μ depends on the particular fluid studied and is called its *viscosity*.

Exercise

1.1   Show that (1.11) is inconsistent with (1.5) unless $\beta - \alpha + 2\mu = 0$.

2   The equations of motion

We continue to assume the fluid is incompressible. Then conservation of mass is represented by the equation

$$\nabla \cdot s = 0. \qquad (2.1)$$

When the reduced stress tensor has the form (1.12), the equation of conservation of momentum (1.7.9) becomes

$$s_t + (s \cdot \nabla) s - \frac{\mu}{\varrho} \nabla^2 s = - \frac{1}{\varrho} \nabla p + Q. \tag{2.2}$$

Equations (2.1) and (2.2) are called the *Navier–Stokes equations*. The quantity μ/ϱ appearing in (2.2) also has a name. μ/ϱ is called the *kinematic viscosity*; it is usually denoted by ν. It should be noted that when $\nu = 0$, the Navier–Stokes equations reduce to the equations (2.1.1–2) of ideal fluid flow.

The Navier–Stokes equations are to be solved along with certain subsidiary conditions. Certainly at any rigid wall, we want the normal component of the velocity to be zero, since this is implied by the condition of specular reflection (theorem 2.1.1). However, this condition is not enough to define a solution of (2.1) and (2.2) uniquely. In a viscous fluid, the molecular attraction between the molecules of the fluid immediately adjacent to a wall and those of the wall itself cause the fluid to adhere to the wall. Therefore, we take as the boundary condition for the velocity s:

$$s = 0 \quad \text{at a rigid boundary.} \tag{2.3}$$

Notice that (2.3) does not have the status of the earlier condition (1.1.3) we used for ideal fluids. (1.1.3) was derived from the condition that the particles making up the fluid reflect specularly from a rigid boundary. (2.3), on the other hand, was not derived from any condition on the system of particles. Moreover, no such condition implying (2.3) is known[1]. In any case, the experimentalists assure us that (2.3) is consistent with what they are able to measure. Also, in accord with Occam's razor, (2.3) is the simplest condition one can think of and should be retained unless contradictions appear.

In addition to (2.3), a condition fixing the flow at some time is required. As in the earlier chapters, this can take the form of an initial condition on s, or a condition that the flow be steady, or perhaps that it be periodic, etc.

3   The stream function

The idea of a stream function proved so useful in dealing with ideal fluids, it is worthwhile to examine the utility of this concept in the viscous case. For convenience, in this section we consider two-dimensional flows for which we can take $w = 0$ and u and v independent of z.

---

[1] One cannot assume that the probability density F of chapter one is zero at a wall, for F is constant on trajectories. If F were zero at a wall, then it would have to be zero for any trajectory that touches the wall.

In this case, equation (2.1) becomes

$$u_x + v_y = 0,$$

which implies there is a function ψ (the stream function) in terms of which we have

$$u = \psi_y, \quad v = -\psi_x. \tag{3.1}$$

A computation similar to the one made in § 6.1 can be repeated here to show that ψ satisfies the equation

$$\left(\frac{\partial}{\partial t} + u \frac{\partial}{\partial x} + v \frac{\partial}{\partial y} - \nu \nabla^2 \right) (\nabla^2 \psi) = 0, \tag{3.2}$$

if Q is zero. Equation (3.2) is a generalization of the earlier (6.1.4).

Again as in chapter six, (3.2) is clearly satisfied whenever ψ is a solution of Laplace's equation:

$$\nabla^2 \psi = 0. \tag{3.3}$$

However—and this is the reason why the stream function is much less useful for viscous fluids than it was for ideal fluids—(3.3) and the condition (2.3) generally imply that ψ is identically constant and, therefore, that the fluid is at rest.

To see why, let ψ be a solution of (3.3) in a domain V. According to (2.3) and (3.1), we require

$$\psi_x = \psi_y = 0 \quad \text{on } \partial V. \tag{3.4}$$

(3.4) clearly entails that both the normal and the tangential derivatives of ψ are zero on ∂V. If ∂V is connected, the tangential derivative of ψ can be integrated to find ψ itself. The result, then, is

$$\frac{\partial \psi}{\partial n} = 0 \quad \text{and} \quad \psi = \text{constant on } \partial V; \tag{3.5}$$

here $\partial/\partial n$ denotes the normal derivative. Now, suppose that ∂V is an analytic arc—a straight line segment or a circular arc, for example. The Cauchy–Kowalewsky theorem[1] says that the problem (3.3), (3.5) has a unique analytic solution, at least in a neighborhood of ∂V. Since $\psi = $ constant is both a solution and analytic, this shows that ψ is constant in a neighborhood of ∂V. The argument can now be repeated to show that ψ is constant everywhere in V.

Thus, in the context of viscous flows, if one is to use a stream function, he will have to consider (3.2) instead of (3.3). Since the main point of the earlier

---

[1] See, *e.g.*, R. Courant and D. Hilbert, Methods of mathematical physics, vol. II. Interscience publishers, New York, 1962.

use of the stream function was to reduce the original equations to the *linear* equation (3.3), and since this is no longer possible for viscous fluids, it is usual to deal directly with the Navier–Stokes equations and not to consider the stream function at all.

4 The energy equation

As in § 3 5, the kinetic energy of an incompressible fluid in a domain V is defined to be

$$\frac{1}{2} \varrho \int_V |s|^2 \, dq.$$

To obtain a formal expression for this quantity, take the dot product of both sides of (2.2) with the vector s. The result is

$$\frac{1}{2} \left(\varrho \frac{\partial}{\partial t} - \mu \nabla^2 \right) |s|^2 + \frac{\mu}{2} |\nabla s|^2 + \nabla \cdot \left(\frac{\varrho}{2} |s|^2 + p \right) s = \varrho s \cdot Q. \quad (4.1)$$

Suppose, for simplicity, we consider a bounded domain V. Green's theorem gives

$$\int_V \nabla^2 |s|^2 \, dq = 0,$$

by (2.3), while the divergence theorem gives

$$\int_V \nabla \cdot \left(\frac{1}{2} |s|^2 + \frac{1}{\varrho} p \right) s \, dq = \int_{\partial V} \left(\frac{1}{2} |s|^2 + \frac{1}{\varrho} p \right) (n \cdot s) \, dA = 0,$$

again by (2.3). Therefore, integrating (4.1) over V, we find

$$\frac{\varrho}{2} \frac{d}{dt} \int_V |s|^2 \, dq + \frac{\mu}{2} \int_V |\nabla s|^2 \, dq = \varrho \int_V Q \cdot s \, dq. \quad (4.2)$$

The first term here is the derivative of the kinetic energy. Equation (4.2) is called the *energy equation*.

An important conclusion can be drawn from the energy equation by taking the external force term Q to be zero. In this case, (4.2) becomes

$$\frac{1}{2} \frac{d}{dt} \int_V |s|^2 \, dq + \nu \int_V |\nabla s|^2 \, dq = 0. \quad (4.3)$$

The second term here obviously has the same sign as ν does. Unless the mechanical energy in the system is to increase when there are no external forces, therefore, we must have $\nu \geqslant 0$. Setting $\nu = 0$ reduces the Navier–

Stokes equations to the equations of ideal fluid flow. Therefore, for viscous fluids,

$$v > 0, \qquad\qquad (4.4)$$

and this assumption is always made in studies of viscous fluids.

5 The existence question

We shall see in the next chapter that solving the Navier–Stokes equations explicitly is generally very difficult. It is natural then to ask two basic questions: under reasonable hypotheses, is there a solution, and if there is, is it unique? These questions have been answered, more or less. But there is at least one surprising gap in the theory.

It was shown in 1950 by Eberhard Hopf[1] that there is always what is called a *weak* solution of the Navier–Stokes equations. What this means roughly is the following. The equations can be converted to an integral equation, and his integral equation has a solution. But it is not known in general whether the solution of the integral equation is sufficiently differentiable to be called a solution of the Navier–Stokes equations.

For two-dimensional flows, the solution is always smooth and the Navier–Stokes equations have a unique solution. In three dimensions, the corresponding result is not known in general. What is known is this. The solution of the initial value problem is smooth if the initial data are small enough (in some suitable topology). It is also smooth for sufficiently small values of the time for any (reasonable) choice of initial data. These results were first proved by A. A. Kiselev and O. A. Ladyzhenskaya[2]. They have been improved by a number of authors[3,4,5].

What is more surprising is this. It is not known whether solutions with large initial data are unique! This is related to the smoothness question. The solution is initially smooth and unique. But at the first value of t where it is not smooth (if there is one), uniqueness may also be lost. This uniqueness question is one of the outstanding questions in theoretical hydrodynamics.

---

[1] Über die Anfangswertaufgabe für die hydrodynamischen Grundgleichungen, Math. Nach. 4 (1950) 213–23.

[2] See the book by Ladyzhenskaya, The mathematical theory of viscous, incompressible flow. Gordon and Breach, New York 1963.

[3] Giovanni Prodi, Teoremi di tipo locale per il sistema di Navier–Stokes e stabilità delle soluzioni stazionarie. Rend. Sem. Mat. Univ. Padova (1962) 374–397.

[4] Tosio Kato and Hiroshi Fujita, On the nonstationary Navier–Stokes system. Rend. Sem. Mat. Padova, 32 (1962) 243–260.

[5] Marvin Shinbrot and Shmuel Kaniel, The initial value problem for the Navier–Stokes equations. Archive for rational mechanics and analysis, 21 (1966) 270–285.

Instant fame awaits the person who answers it. (Especially if the answer is negative!)

If it should turn out that solutions are not unique, that fact would qualify as a paradox and would justify complicating the model derived in § 1. There are two obvious ways in which the model can be changed. The first is to allow the reduced stress tensor to depend on higher derivatives of the velocity than the first (cf. (1.1)). The second is to allow the reduced stress tensor to depend nonlinearly on the derivatives of the velocity. A piquant aspect is lent to the whole problem by the fact that in *either* case the solution can be shown to be unique[1]!

It might be argued that as a practical matter the whole question may be irrelevant since solutions are known to be unique if the data are small. It is true, after all, that for large initial data no fluid can be looked upon as incompressible. However, the values of the initial data for which the uniqueness question can be settled are *too* small. They are well below the values for which compressibility effects in water, say, come into effect or in which our hypotheses can be expected to fail. Thus, even from a "practical" point of view, the existence and uniqueness theorems require strengthening.

In any case, the uniqueness question remains an aggravating mathematical problem which seems extremely difficult for no very good reason.

The existence and uniqueness questions are taken up in Part II of this book.

---

[1] For the first, see Ladyzhenskaya, *op. cit.*, p. 159, for the second, Shmuel Kaniel, On the motion of a viscous incompressible fluid. Journal of mathematics and mechanics 19 (1970) 681–707.

CHAPTER 8

Examples of viscous fluid flow

1   Introduction

THE NAVIER–STOKES equations are nonlinear and, as we saw in § 7.3, they cannot be reduced to linear equations, at least not by the simple scheme of introducing a stream function that worked so well for ideal fluids. A consequence of the nonlinearity of the equations is that there are only a few exact solutions known. An interesting sidelight is that for most of the known solutions, the nonlinear term $(s \cdot \nabla) s$ vanishes identically.

We present five examples. In four of them the geometry is such that there is a solution for which $(s \cdot \nabla) s$ vanishes. The fifth displays very well the astonishing complication of which nonlinear flows are capable.

In all the examples, we take the external force to be zero. Then, the Navier–Stokes equations become

$$s_t + (s \cdot \nabla) s - \nabla^2 s = - \frac{1}{\varrho} \nabla p, \qquad (1.1)$$

$$\nabla \cdot s = 0 \qquad (1.2)$$

2   Steady flow between parallel planes

Our first example is one of two-dimensional, steady flow between two fixed, parallel planes. As usual for two-dimensional flows, we choose the coordinate system such that $w = 0$, while u, v, and p are independent of z. Since the flow is steady, they are also independent of t.

Let the planes be represented by the lines $y = \pm h$. We ask if there is a flow between these lines for which $v \equiv 0$. If there is, then the mass conservation equation (1.2) gives

$$u_x = 0, \qquad (2.1)$$

so that u is independent of x. In that case, equations (1.1) become linear:

$$-v u_{yy} + \frac{1}{\varrho} p_x = 0, \qquad (2.2)$$

$$p_y = 0. \qquad (2.3)$$

(2.3) says that p can only depend on x. But (2.2) and (2.1) say that p_x is a function only of y. Therefore, p_x must be constant. In terms of this constant,

8   Shinbrot (0171)

(2.2) can be integrated to give (recall $v = \mu/\varrho$)

$$u = \alpha + \beta y + \frac{p_x}{2\mu} y^2,$$

where α and β are integration constants.

Since u is to be zero on the walls $y = \pm h$ (cf. (7.2.3)), α and β can be evaluated. We find

$$u = -\frac{p_x}{2\mu} (h^2 - y^2).$$

The velocity field for fixed x is depicted in Figure 25. In the figure, each arrow represents the velocity of the fluid at the point on the base of the arrow. The heads of the arrows lie on a parabola.

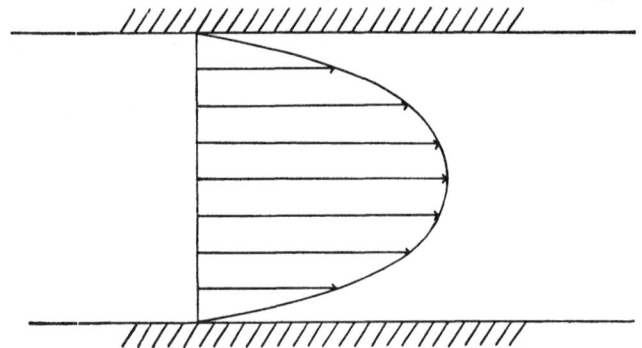

Figure 25

3   Steady flow in a pipe

We consider next steady flow in a cylinder of arbitrary cross section; we call such a cylinder a *pipe*. Take the x-axis parallel to the generators of the pipe. This time, we ask if there is a flow in which v and w are zero, while u depends only on y and z. Clearly, in such a case, (1.2) is satisfied automatically. The components of (1.1) in the y- and z-directions give, as in § 2, that p is a function of x alone. The x-component of (1.1) is

$$u_{yy} + u_{zz} = \frac{1}{\mu} p_x. \tag{3.1}$$

The left side of this equation is independent of x, while the right side depends only on x. Therefore, p_x is constant, and the problem reduces to solving *Poisson's equation* (3.1) with constant right hand side.

A simple case occurs when the pipe has circular cross-section, of radius R, say. Write

$$y = r \cos \theta$$

$$z = r \sin \theta,$$

and suppose that u depends only on r. Then (3.1) becomes

$$u_{rr} + \frac{1}{r} u_r = \frac{1}{\mu} p_x.$$

The general solution of this equation is

$$u = \alpha + \beta \log r + \frac{p_x r^2}{4\mu};$$

here, α and β are constants.

If the velocity is everywhere finite in the flow, $\beta = 0$. Since $u = 0$ on the walls of the pipe when $r = R$, we have

$$u = -\frac{p_x}{4\mu} (R^2 - r^2). \tag{3.2}$$

As in the flow of § 2, the speed varies parabolically across the pipe and has a maximum when $r = 0$. The flow described by (3.2) is called *Poiseuille flow*.

4   Steady flow past a moving plane

In this section, we consider a flow past a *moving* boundary. For such flows, the boundary condition (7.2.3) must be changed. (7.2.3) was derived from the hypothesis that the fluid at a wall sticks to it. Under the same hypothesis, the velocity of a fluid at a moving wall must be the same as the velocity of the wall itself. In this section, we take the wall to be a plane moving with constant velocity in its own plane. Once again, we ask if there is a steady, two-dimensional flow in a half-space bounded by such a plane. As usual, then, we take $w = 0$ and u, v, and p independent of z. We take the plane to be $y = 0$. As in § 2, the Navier–Stokes equations (1.1–2) become

$$u_x = 0,$$

$$p_y = 0,$$

$$u_{yy} = \frac{1}{\mu} p_x.$$

Once again, these equations imply that p_x is constant. Therefore,

$$u = \alpha + \beta y + \frac{p_x}{2\mu} y^2,$$

α and β being integration constants.

Let the plane $y = 0$ move with speed u_0 to the right. Then, when $y = 0$, we required $u = u_0$. Therefore

$$u = u_0 + \beta y + \frac{p_x}{2\mu} y^2.$$

If we require that the speed of the fluid be bounded, we see that we must have $p_x = 0$, and we must choose $\beta = 0$. Thus, we have proved that the only two-dimensional, steady flow past a plane moving with the velocity u_0 with the velocity everywhere bounded is the trivial one $u = u_0$.

5   Unsteady flow past a moving plane

A slightly different problem is obtained by letting the fluid be at rest initially, and at time zero to start the plane $y = 0$ moving with constant velocity u_0 to the right. This is a non-steady problem which differs, therefore, from the others we have considered.

We assume the problem is two-dimensional and that there is no flow in the y-direction. Then, $v = w = 0$, and once again (1.2) gives

$$u_x = 0.$$

The component of (1.1) in the y-direction is

$$p_y = 0,$$

and that in the x-direction is

$$u_t - \nu u_{yy} = -\frac{1}{\varrho} p_x.$$

The left-hand side of this equation depends on y and on t; the right side depends on x and on t. Thus, p_x is a function of t alone. We show there actually is a solution in which the pressure is identically constant.

When p is constant, $u = u(t, y)$ satisfies

$$u_t = \nu u_{yy}, \tag{5.1}$$

$$u(0, y) = 0, \tag{5.2}$$

$$u(t, 0) = u_0. \tag{5.3}$$

Define the *Laplace transform* of u by

$$\hat{u}(\tau, y) = \int_0^\infty e^{-\tau t} u(t, y) \, dt. \tag{5.4}$$

It is easy to see that if u satisfies (5.2), the Laplace transform of u_t is just $\tau \hat{u}\,(\tau, y)$. Formally, then, $\hat{u}$ satisfies

$$\hat{u}_{yy} - \frac{\tau}{\nu}\, \hat{u} = 0 \tag{5.5}$$

$$\hat{u}\,(\tau, 0) = \frac{u_0}{\tau}. \tag{5.6}$$

The general solution of the ordinary differential equation (5.5) is

$$\hat{u}\,(\tau, y) = \alpha(\tau)\, e^{-y\sqrt{\tau/\nu}} + \beta(\tau)\, e^{y\sqrt{\tau/\nu}}.$$

We take the domain of the fluid to be the half-plane $y > 0$. Clearly, (5.4) only makes sense if Re $\tau > 0$. If $\hat{u}\,(\tau, y)$ (and, therefore, $u\,(t, y)$) is to be bounded as $y \to \infty$, we must choose $\beta(\tau) \equiv 0$. Imposing the initial condition (5.6), then, we find

$$\hat{u}\,(\tau, y) = \frac{u_0}{\tau}\, e^{-y\sqrt{\tau/\nu}}.$$

The inverse transform of this function is known[1]; it is

$$u\,(t, y) = \frac{2u_0}{\sqrt{\pi}} \int_{y/(2\sqrt{\nu t})}^{\infty} e^{-\lambda^2}\, d\lambda. \tag{5.7}$$

It can be verified by substitution that (5.7) satisfies (5.1–3). To do so, one only needs a single fact, and that is that

$$\int_0^\infty e^{-\lambda^2}\, d\lambda = \frac{\sqrt{\pi}}{2}.$$

Another remark about this problem is that as $t \to \infty$, $y/(2\sqrt{\nu t}) \to 0$, so that $u\,(t, y) \to u_0$, the solution of the steady problem of § 4.

Finally, notice the moving boundary $y = 0$ pulls the fluid along with it, which could never happen if the flow were ideal. This illustrates the fact that we have been successful in portraying frictional forces by the simple form (7.1.12) of the Stokes hypothesis.

6   Viscous flow due to a source

In all the preceding examples, the nonlinear term $(s \cdot \nabla)\, s$ vanished identically. That was not an accident; exact solutions of the full nonlinear equations are rare. We consider here an example of such a solution: it describes the steady two-dimensional flow between two plane walls meeting at an angle 2α.

---

[1] See A. Erdelyi, *et al.* Tables of integral transforms, volume 1. McGraw-Hill, 1954.

The point of intersection of the planes is taken to be either a source or a sink[1].

Introduce polar coordinates with the origin at the source and the walls at $\theta = \pm\alpha$. If R and Θ denote the radial and angular components of the velocity, then a straightforward calculation shows that the two-dimensional, steady Navier–Stokes equations take the form

$$RR_r + \frac{1}{r}\,\Theta R_\theta - \frac{1}{r}\,\Theta^2 = -\frac{1}{\varrho}\,p_r + \nu\left(R_{rr} + \frac{1}{r^2}\,R_{\theta\theta} + \frac{1}{r}\,R_r - \frac{2}{r^2}\,\Theta_\theta - \frac{1}{r^2}\,R\right),$$

$$R\Theta_r + \frac{1}{r}\,\Theta\Theta_\theta + \frac{1}{r}\,R\Theta = -\frac{1}{\varrho r}\,p_\theta + \nu\left(\Theta_{rr} + \frac{1}{r^2}\,\Theta_{\theta\theta} + \frac{1}{r}\,\Theta_r + \frac{2}{r^2}\,R_\theta - \frac{1}{r^2}\,\Theta\right),$$

$$R_r + \frac{1}{r}\,R + \frac{1}{r}\,\Theta_\theta = 0.$$

In the problem we are considering, it is appropriate to assume the flow is purely radial. (We shall have complications enough.) Setting $\Theta = 0$, then, the above equations become

$$RR_r + \frac{1}{\varrho}\,p_r = \nu\left(R_{rr} + \frac{1}{r^2}\,R_{\theta\theta} + \frac{1}{r}\,R_r - \frac{1}{r^2}\,R\right), \tag{6.1}$$

$$\frac{1}{\varrho}\,p_\theta = \frac{2\nu}{r}\,R_\theta, \tag{6.2}$$

$$\frac{\partial}{\partial r}\,(rR) = 0. \tag{6.3}$$

The last of these equations shows that rR is a function of θ only. Define

$$S(\theta) = \pm rR\,(r, \theta),$$

where we take the plus sign if the origin is a source, and the minus sign if it is a sink. The reason for the variation in sign is, of course, the fact that we expect R to be generally positive if we have a source, and generally negative if we have a sink.

In terms of S, equations (6.1) and (6.2) become

$$\frac{1}{\varrho}\,p_r = \pm\frac{1}{r^3}\,(\nu S_{\theta\theta} \pm S^2)$$

$$\frac{1}{\varrho}\,p_\theta = \pm\frac{2\nu}{r^2}\,S_\theta.$$

---

[1] This flow was first discussed by G. Hamel, Spiralförmige Bewegung zäher Flüssigkeiten. Jahres. der Deutschen Mathematiker-Vereinigung, 25 (1916) 34–60.

p can be eliminated from these equations by differentiating the first with respect to θ, the second with respect to r, and subtracting the results. The result is

$$\frac{1}{r^3}\frac{d}{d\theta}(\nu S_{\theta\theta} \pm S^2 + 4\nu S) = 0.$$

Thus, with a denoting a constant, we have

$$\nu S_{\theta\theta} \pm S^2 + 4\nu S = \frac{a}{2}. \tag{6.4}$$

This equation can be solved for θ as a function of S. Multiply (6.4) by S_θ and integrate with respect to θ. The result is

$$\frac{\nu}{2}S_\theta^2 \pm \frac{1}{3}S^3 + 2\nu S^2 = \frac{a}{2}S + \frac{b}{2},$$

b being another constant. Therefore, we have

$$\theta = c \pm \sqrt{\nu}\int \frac{dS}{\sqrt{b + aS - 4\nu S^2 \mp \frac{2}{3}S^3}}, \tag{6.5}$$

c being one final constant, and the additional ambiguity in sign being due to the square root.

There are three undetermined constants in (6.5). Two of them will be determined by the condition that the velocity be zero at the walls. The third is determined by the strength of the source at the origin.

The condition that the origin is a source amounts to the hypothesis that the mass flux[1] through any curve $r = $ constant is constant. In symbols, if M denotes the mass flux,

$$M = \varrho \int_{-\alpha}^{\alpha} Rr\, d\theta$$

$$= \pm\varrho \int_{-\alpha}^{\alpha} S(\theta)\, d\theta. \tag{6.6}$$

It is important to notice that M can be either positive or negative. M is the mass of fluid per unit time crossing a curve $r = $ constant in the direction of r increasing. Therefore, if M is positive, the origin is a source, but if one wishes the origin to be a sink, he must take M negative.

We consider first the case $M < 0$, since the situation there is simpler than it is when $M > 0$. This means that we must choose the lower sign in all of the above calculations. We ask whether there can be a flow which is symmetric

---

[1] That is, the mass per unit time.

in θ and in which the speed increases monotonically from the value zero at the walls to a maximum when $\theta = 0$.

If such a flow exists, it must be true that $S_\theta(0) = 0$, since $\theta = 0$ is a maximum of $S(\theta)$. Therefore, $S(0)$ must be a zero of the cubic occurring in the denominator of (6.5), and we can write

$$b + aS - 4\nu S^2 + \tfrac{2}{3}S^3 = [S(0) - S][\sigma + \{4\nu - \tfrac{2}{3}S(0)\} S - \tfrac{2}{3}S^2],$$

σ being another constant. Since $S(0)$ is, by definition, the value of S when $\theta = 0$, we obtain from (6.5) that

$$\theta = \pm\sqrt{\nu} \int_{S(0)}^{S} \frac{dS}{\sqrt{[S(0) - S][\sigma + \{4\nu - \tfrac{2}{3}S(0)\} S - \tfrac{2}{3}S^2]}}.$$

In the type of flow we are considering, $S \leqslant S(0)$. Therefore, the integral here is non-positive. If θ is to vary from 0 to λ, then, we must choose the lower sign to obtain, finally,

$$\theta = \sqrt{\nu} \int_{S}^{S(0)} \frac{dS}{\sqrt{[S(0) - S][\sigma + \{4\nu - \tfrac{2}{3}S(0)\} S - \tfrac{2}{3}S^2]}} \qquad (6.7)$$

The constants $S(0)$ and σ remain to be determined. The conditions that determine them are

$$\alpha = \sqrt{\nu} \int_{0}^{S(0)} \frac{dS}{\sqrt{[S(0) - S][\sigma + \{4\nu - \tfrac{2}{3}S(0)\} S - \tfrac{2}{3}S^2]}},$$

and

$$\tfrac{1}{2}|M| = \varrho \int_{0}^{\lambda} S(\theta)\, d\theta$$

$$= \varrho \int_{S(0)}^{0} S \frac{d\theta}{dS}\, dS$$

$$= \varrho \sqrt{\nu} \int_{0}^{S(0)} \frac{S\, dS}{\sqrt{[S(0) - S][\sigma + \{4\nu - \tfrac{2}{3}S(0)\} S - \tfrac{2}{3}S^2}}.$$

It is convenient to transform these equations by writing $S = S(0)\lambda, \sigma = S(0)\tau$. We then obtain

$$\lambda = \sqrt{\nu} \int_{0}^{1} \frac{d\lambda}{\sqrt{(1 - \lambda)[\tau + \{4\nu - \tfrac{2}{3}S(0)\} \lambda - \tfrac{2}{3}S(0) \lambda^2}}, \qquad (6.8)$$

$$\tfrac{1}{2}|M| = \varrho \sqrt{\nu}\, S(0) \int_{0}^{1} \frac{\lambda\, d\lambda}{\sqrt{(1 - \lambda)[\tau + \{4\nu - \tfrac{2}{3}S(0)\} \lambda - \tfrac{2}{3}S(0) \lambda^2}}.$$

$$(6.9)$$

In order for the integrands here to be real, it is necessary that

$$\tau \geqslant 0, \tag{6.10}$$

$$\tau + 4\nu \geqslant \tfrac{4}{3}S(0). \tag{6.11}$$

(6.8) defines τ as a function of $S(0)$ for any $S(0) \geqslant 0$. To see this, suppose first that $S(0) \geqslant 3\nu$. Then, (6.11) implies (6.10). Moreover, as $\tau \to \infty$, the right side of (6.8) goes to zero, while as $\tau \to \tfrac{4}{3}S(0) - 4\nu$, the right side goes to infinity. Thus, for any value of $S(0) \geqslant 3\nu$, there is at least one value of τ that validates (6.8).

If $0 \leqslant S(0) < 3\nu$, (6.10) implies (6.11). Again as $\tau \to \infty$, the right side of (6.8) goes to zero. At the other extreme, when $\tau \to 0$, the right side of (6.8) goes to

$$\sqrt{\nu} \int_0^1 \frac{d\lambda}{\sqrt{\lambda\,(1-\lambda)\,[4\nu - \tfrac{4}{3}S(0)\,(1+\lambda)]}}. \tag{6.12}$$

This integral is an increasing function of $S(0)$, and for $S(0) = 0$, its value is

$$\frac{1}{2} \int_0^1 \frac{d\lambda}{\sqrt{\lambda\,(1-\lambda)}} = \frac{\pi}{2}.$$

Thus, (6.12) exceeds $\pi/2$ for *every* value of $S(0)$, so that whenever $\alpha \leqslant \pi/2$, there is still at least one value of τ validating (6.8). Thus, we see that this is true for all $S(0) \geqslant 0$.

Moreover, with α fixed, there is only one τ for which (6.8) holds for any given $S(0)$. This is so because, differentiating (6.8) implicitly with respect to $S(0)$, we find

$$\frac{d\tau}{dS(0)} = \frac{2}{3} \frac{\displaystyle\int_0^1 \frac{\lambda\,(1-\lambda^2)\,d\lambda}{\{(1-\lambda)\,[\tau + 4\lambda - \tfrac{4}{3}S(0)\,\lambda\,(1+\lambda)]\}^{3/2}}}{\displaystyle\int_0^1 \frac{(1-\lambda)\,d\lambda}{\{(1-\lambda)\,[\tau + 4\lambda - \tfrac{4}{3}S(0)\,\lambda\,(1+\lambda)]\}^{3/2}}}, \tag{6.13}$$

and this is positive. Thus, τ is a monotonically increasing function of $S(0)$.

We now consider (6.9). Since τ increases with $S(0)$, (6.11) shows that $\tau \to \infty$ as $S(0) \to \infty$. Moreover, a quick calculation with (6.8) shows that $\tau = O\,(S(0))$ as $S(0) \to \infty$. Therefore, the right side of (6.9) approaches infinity like $[S(0)]^{1/2}$ as $S(0) \to \infty$. Clearly, the right side of (6.9) goes to zero as $S(0)$ does. Therefore, there is a value of $S(0)$ such that (6.9) is valid.

We conclude, then, that for every angle $\alpha \leqslant \pi/2$, and for every negative value of the flux M, there is a solution of our problem with the flow purely radial, with the radial speed symmetric about $\theta = 0$ and increasing monotonically as θ goes from α to 0. A profile of the flow is sketched in Figure 26.

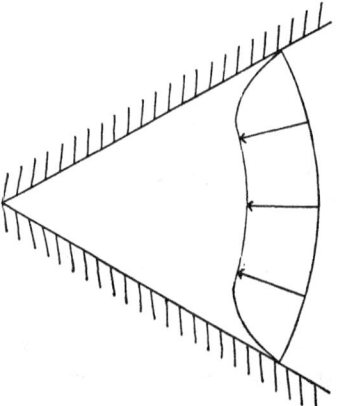

Figure 26

Now, suppose that M is positive. If the preceding calculations are repeated with the appropriate changes, one finds that the equations equivalent to (6.8) and (6.9) that we must solve are

$$\alpha = \sqrt{\nu} \int_0^1 \frac{d\lambda}{\sqrt{(1 - \lambda)\,[\tau + \{4\nu + \tfrac{2}{3}S(0)\}\,\lambda + \tfrac{2}{3}S(0)\,\lambda^2]}}, \qquad (6.14)$$

$$\tfrac{1}{2}M = \varrho\,\sqrt{\nu}\,S(0) \int_0^1 \frac{\lambda\,d\lambda}{\sqrt{(1 - \lambda)\,[\tau + \{4\nu + \tfrac{2}{3}S(0)\}\,\lambda + \tfrac{2}{3}S(0)\,\lambda^2]}}. \qquad (6.15)$$

Suppose that for some value of α, we think of (6.14) as defining $S(0)$ as a function of τ. A calculation similar to the one that led to (6.13) shows that in this case, $dS(0)/d\tau$ is negative. Thus, $S(0)$ is a decreasing function of τ, and for a given α, the greatest value of $S(0)$ that can be obtained is the one computed from (6.14) with $\tau = 0$. We denote this maximum value of $S(0)$ by $S_m = S_m(\alpha)$. It satisfies

$$\alpha = \sqrt{\nu} \int_0^1 \frac{d\lambda}{\sqrt{\lambda\,(1 - \lambda)\,[4\nu + \tfrac{2}{3}S_m\,(1 + \lambda)]}}.$$

(6.15) is obviously a decreasing function of τ, and differentiation of (6.15) shows that it is an increasing function of $S(0)$. Therefore, the maximum attainable value of M does not exceed the right side of (6.15) with $\tau = 0$ and $S(0) = S_m(\alpha)$:

$$\tfrac{1}{2}M \leqslant \varrho\,\sqrt{\nu}\,S_m \int_0^1 \frac{\lambda\,d\lambda}{\sqrt{\lambda\,(1 - \lambda)\,[4\nu + \tfrac{2}{3}S_m\,(1 + \lambda)]}}.$$

The conclusion we draw is that for each α, there is a *maximum* mass flux for which a symmetric flow of the type we have been considering is possible. A flow of this type is sketched in Figure 27. If the mass flux exceeds the critical

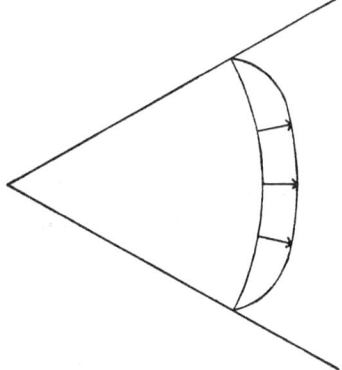

Figure 27

value, it turns out that the flow is not everywhere directed away from the source, but is like one of the flows sketched in Figure 28.

As M increases, the number of sign changes suffered by the velocity is forced to increase. Presumably, this fact indicates the onset of turbulence.

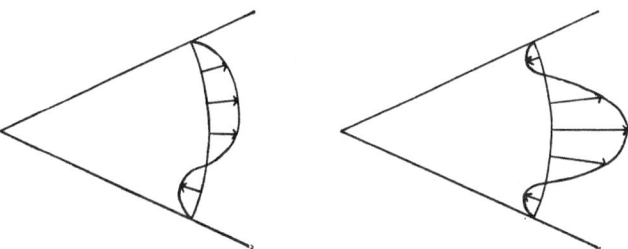

Figure 28

Exercises

6.1 Prove in detail that for two-dimensional flow due to a source at the intersection of two planes, there is a maximum mass flux for which the flow is symmetric, everywhere radial, and everywhere directed away from the source.

6.2 Show that in three-dimensional viscous flow in a cone with a source (or a sink) at its apex, there is never a steady flow where the velocity is purely radial. Can steady flow in such a cone exist (without the assumption that the flow is radial)?

CHAPTER 9

Various approximations

1 Introduction

PERHAPS THE most important conclusion that can be drawn from the last chapter (particularly § 8.6) is that solutions of the Navier–Stokes equations, even in simple situations, can be *complicated*. The difficulties notwithstanding, the engineer must have some estimate for, say, the forces on a body in a viscous fluid. Therefore, various approximate schemes have been suggested for solving the equations. It is mostly these approximate equations that are the source of the spate of nomenclature that occurs in incompressible fluid theory and that so confuses the mathematician wishing to enter the field.

We consider three sets of approximate equations here. They are the three most commonly used in viscous flow theory.

2 Stokes flow

The steady Navier–Stokes equations with no external forces are

$$(s \cdot \nabla) s - \nu \nabla^2 s = -\frac{1}{\varrho} \nabla p, \qquad (2.1)$$

$$\nabla \cdot s = 0. \qquad (2.2)$$

It was suggested by Stokes that if the velocity is small enough, the quadratic term $(s \cdot \nabla) s$ is small and can be ignored. In that case, (2.1) becomes

$$\nabla^2 s = \frac{1}{\mu} \nabla p. \qquad (2.3)$$

This is to be solved along with the continuity equation (2.2) and appropriate boundary conditions. Any flow for which the associated velocity vector satisfies (2.2) and (2.3) is called *Stokes flow* (it is also called *creeping flow*), and we refer to (2.2) and (2.3) as the *Stokes equations*. It should be noted that they have the inestimable advantage over the Navier–Stokes equations of being linear; I do not know any other virtues that they possess.

Naturally, every example in chapter eight for which the nonlinearity vanishes is an example of Stokes flow. We look at some other examples here. (1) Consider the Stokes equations for flow past a sphere. Then, in spherical polar coordinates[1] (r, θ, ω), the boundary conditions are

$$s = 0 \quad \text{for} \quad r = R \quad \text{(say)} \tag{2.4}$$

and

$$s \to U \quad \text{as} \quad r \to \infty, \tag{2.5}$$

where U is a constant vector which we may take to be parallel to the x-axis.

It is convenient in this case to suppose that the fluid is stationary at infinity and that the sphere moves through it with velocity $- U$. Formally, this change is effected by defining

$$\sigma = s - U. \tag{2.6}$$

Then, the vector σ satisfies the Stokes equations (2.2) and (2.3), the condition

$$\sigma \to 0 \quad \text{as} \quad r \to \infty, \tag{2.7}$$

and the boundary condition

$$\sigma = -U \quad \text{when} \quad r = R. \tag{2.8}$$

In two dimensions, the continuity equation (2.2) implies the existence of a stream function; in three dimensions, the continuity equation implies that there is a vector Ψ such that

$$\sigma = \nabla \times \Psi. \tag{2.9}$$

Writing σ in this form implies automatically that it satisfies (2.2). Substituting (2.9) into (2.3), we obtain

$$\nabla^2 (\nabla \times \Psi) = \frac{1}{\mu} \nabla p$$

or, since $\nabla \times \nabla p = 0$,

$$\nabla^2 (\nabla \times \nabla \times \Psi) = 0. \tag{2.10}$$

We attempt to find Ψ in the form

$$\Psi = \nabla \psi \times U = \nabla \times (\psi U), \tag{2.11}$$

where ψ is a scalar function of r alone. (2.10) gives

$$\nabla^2 (\nabla \times \nabla \times \nabla \times \psi U) = 0.$$

But

$$\nabla \times \nabla \times \psi U = \nabla (\nabla \cdot \psi U) - \nabla^2 (\psi U),$$

so that

$$\nabla \times \nabla \times \nabla \times \psi U = -\nabla^2 (\nabla \times \psi U).$$

---

[1] We use the notation of § 2.4 for spherical polar coordinates.

Thus,

$$\nabla^2 (\nabla \times \nabla \times \nabla \times \psi U) = -\nabla^4 (\nabla \times \psi U) = -\nabla^4 (\nabla \psi \times U)$$

$$= -\nabla^4 (\nabla \psi) \times U = 0.$$

This means that $\nabla^4 (\nabla \psi) = \nabla (\nabla^4 \psi)$ is parallel to U:

$$\nabla (\nabla^4 \psi) = f(q, \psi) \, U,$$

where f is a scalar. We try to find a solution for which[1] $f = 0$. Then $\nabla^4 \psi$ is constant. We take the condition (2.7) at infinity to mean[1] that all derivatives of ψ of order two or higher are zero at infinity. Letting $r \to \infty$ in the equation $\nabla^4 \psi = $ constant, then, we find that the constant is zero, and

$$\nabla^4 \psi = 0.$$

Since ψ depends on r alone, this equation takes the form

$$\frac{1}{r^2} \frac{d}{dr} \left(r^2 \frac{d}{dr} \right) \nabla^2 \psi = 0,$$

so that

$$\nabla^2 \psi = \frac{\alpha}{r} + \beta,$$

where α and β are constants. Letting $r \to \infty$, we get $\beta = 0$. Therefore,

$$\frac{1}{r^2} \frac{d}{dr} \left(r^2 \frac{d}{dr} \right) \psi = \frac{\alpha}{r},$$

and

$$\psi = \frac{\alpha r}{2} + \frac{\gamma}{r}, \tag{2.13}$$

γ being another constant. Since the velocity depends only on *derivatives* of ψ (see (2.9)), a final additive constant has been set equal to zero in determining (2.13).

Whatever values we choose for the constants α and γ, the associated velocity, obtained through (2.9) and (2.11), will satisfy (2.2), (2.3), and (2.7). α and γ must be chosen so that (2.8) is satisfied. Now, (2.13), along with (2.9) and (2.11), show that

$$\sigma = \nabla \times \nabla \times (\psi U) = -\nabla^2 (\psi U) + \nabla (\nabla \cdot \psi U)$$

$$= -\left(\frac{\alpha}{2r} + \frac{\gamma}{r^3} \right) U - \left(\frac{\alpha}{2r^3} - \frac{3\gamma}{r^5} \right) (U \cdot q) \, q.$$

---

[1] It can be proved that the solution to the linear problem (2.2–5) is unique. Therefore, any solution that we find, by means of whatever bizarre assumptions, is *the* solution.

When $r = R$, we must have

$$-U = -\left(\frac{\alpha}{2R} + \frac{\gamma}{R^3}\right) U - \left(\frac{\alpha}{2R^3} - \frac{3\gamma}{R^5}\right)(U \cdot q)\, q.$$

This must hold for all vectors q of length R. Therefore,

$$\frac{\alpha}{2R} + \frac{\gamma}{R^3} = 1$$

$$\frac{\alpha}{2R^3} - \frac{3\gamma}{R^5} = 0,$$

or $\alpha = 3R/2$, $\gamma = R^3/4$. The velocity, then, is

$$\sigma = -\left(\frac{3R}{4r} + \frac{R^3}{4r^3}\right) U + \left(\frac{3R^3}{4r^5} - \frac{3R}{4r^3}\right)(U \cdot q)\, q. \qquad (2.14)$$

The streamlines of the flow are sketched in Figure 29[1]. The fact that the flow does not look like what one would expect is discussed further in § 3.

Now that we know the velocity, the pressure can be computed from (2.2) and from that, the force. It is a worthwhile exercise to evaluate the force on a sphere in Stokes flow. The result is

$$F = 6\pi R\mu\, U. \qquad (2.15)$$

This result is called the *Stokes formula*. It should be observed that this force is in the direction of the velocity at infinity and so is a pure drag; there is no lift. (2.15) has been verified experimentally for very small velocities.

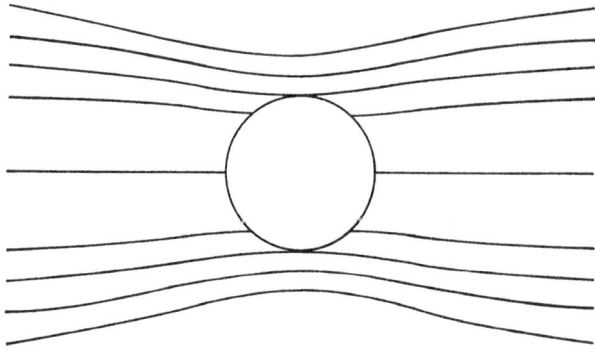

Figure 29

(2) Analogous to the flow around a sphere discussed in example (1) is the

---

[1] Figure 29 is taken from Lamb, Hydrodynamics. New York, Dover, 1945.

two-dimensional Stokes flow past a circular cylinder. To find this flow, we assume the fluid is stationary at infinity and that the cylinder moves with velocity $- U$ in the direction of the x-axis. Then, we must solve the equations

$$u_{xx} + u_{yy} = \frac{1}{\mu} p_x, \tag{2.16}$$

$$v_{xx} + v_{yy} = \frac{1}{\mu} p_y, \tag{2.17}$$

$$u_x + v_y = 0, \tag{2.18}$$

with the boundary conditions

$$u = - U, \quad v = 0 \quad \text{when} \quad r = R, \tag{2.19}$$

and the condition at infinity

$$u \to 0, \quad v \to 0 \quad \text{as} \quad r \to \infty. \tag{2.20}$$

Let ψ be the stream function associated with the assumed flow. Then, just as in example (1), ψ satisfies

$$\nabla^4 \psi = 0. \tag{2.21}$$

We suppose that ψ depends on r alone. In two dimensions, then, (2.21) becomes

$$\frac{1}{r} \frac{d}{dr} \left(r \frac{d}{dr} \right) \nabla^2 \psi = 0.$$

Therefore,

$$\nabla^2 \psi = \alpha \log r + \beta, \tag{2.22}$$

where α and β are constants. We do not assume in this case that $\nabla^2 \psi \to 0$ as $r \to \infty$, but we integrate (2.22) as it stands. The result then is

$$\psi_r = \alpha r \left(\frac{1}{2} \log r - \frac{1}{4} \right) + \frac{\beta r^2}{3} + \frac{\gamma}{r},$$

where γ is another integration constant.

We now use (2.20) and infer that α and β are zero. Therefore,

$$\psi_r = \frac{\gamma}{r}.$$

To impose (2.19), we compute v when $r = R$:

$$v = -\psi_x = -\psi_r \cos \theta = - \frac{\gamma}{R} \cos \theta.$$

If this is to be zero, γ must be zero, and then u and v are identically zero! Therefore, there is no flow of the type we have been trying to find.

This is the *Stokes paradox*: there is no steady two-dimensional Stokes flow past a cylinder with constant, non-zero velocity at infinity. We have actually proved a weak version of the paradox, since we have assumed that ψ depends only on r, but knowing what to expect, it is not hard to prove the full theorem.

It can be proved that the steady Navier–Stokes equations do have a solution in the same circumstances. What the Stokes paradox means, then, is that however small the speed may be at infinity. *the nonlinear term $(s \cdot \nabla) s$ in the Navier–Stokes equations can never be small* compared with the other terms *uniformly in the entire domain of the fluid for flow past a cylinder*. In turn, this can be interpreted as meaning that the Navier–Stokes equations are *essentially* nonlinear, and the hard part of the problem can never be wished away.

Exercise

2.1   Prove the full Stokes paradox: The two-dimensional equations

$$\nabla^2 s = \frac{1}{\mu} \, \nabla p$$

$$\nabla \cdot s = 0$$

do not have a solution in the exterior of the disk $|q| > R$ satisfying

$$s = -U \quad \text{when} \quad |q| = R$$

$$s \to 0 \quad \text{as} \quad |q| \to \infty$$

and

$$\int_{|q| > R} |\nabla s|^2 \, dq < \infty.$$

3   Oseen flow

The concluding remarks of the last section are reinforced by the following argument, due essentially to Oseen[1]. Consider a three-dimensional flow past a body—say a sphere. We suppose that the velocity goes to the constant velocity U at infinity. If the Stokes flow computed in § 2(1) gives an accurate approximation to the solution of the Navier–Stokes equations, the term $(s \cdot \nabla) s$ in these equations must be small compared, say, with the term $\nu \nabla^2 s$.

Now, for large r, u is approximately U, while for Stokes flow, (2.14) shows that the first derivatives of u (or v, they are the same) go to zero like $1/r^2$, at best. Thus, the term $(s \cdot \nabla) s$ is $0 \, (1/r^2)$ as $r \to \infty$, and no smaller.

Again using (2.14), we find on the other hand that the second derivatives

---

[1] Über die Stokes'sche Formel, und über eine verwandte Aufgabe in der Hydrodynamik. Arkiv für Mat., 6 (1910).

of u are $0\,(1/r^3)$ at least. In fact, $\nabla^2 u$ is smaller than that since the largest term in (2.14)—the term $(3R/4r)\,U$—has vanishing Laplacian. Therefore, $\nabla^2 u = 0\,(1/r^5)$ as $r \to \infty$. Consequently, except for certain special directions (possibly), the ratio of $|(s \cdot \nabla)\,s|$ to $|\nabla^2 s|$ *grows* like r^3 as $r \to \infty$. Therefore, even in three-dimensional flow the nonlinear term cannot be ignored throughout the domain of the fluid, and the Stokes equations cannot be expected to be uniformly valid throughout the fluid. This fact is the cause of the strangeness of Figure 29.

In addition to this criticism, Oseen also suggested a remedy. Since the difficulty with Stokes flow occurs at infinity, and since the velocity of the fluid is known at infinity, Oseen suggested writing

$$s = U + \sigma \tag{3.1}$$

in the Navier–Stokes equations and ignoring the quadratic term $(\sigma \cdot \nabla)\,\sigma$ that occurs there and that *will* be small at infinity.

If the substitution (3.1) is made in the steady Navier–Stokes equations (2.1) and the term $(\sigma \cdot \nabla)\,\sigma$ stricken from the resulting equations, one obtains

$$-\nu\,\nabla^2\sigma + (U \cdot \nabla)\,\sigma = -\frac{1}{\varrho}\,\nabla p. \tag{3.2}$$

This is known as *the Oseen equation*. It is a linearized model of the Navier–Stokes equations and must be solved along with the equation of continuity

$$\nabla \cdot \sigma = 0, \tag{3.3}$$

the boundary condition

$$\sigma = -U \tag{3.4}$$

on any bodies immersed in the flow, and the condition

$$\sigma \to 0 \quad \text{at infinity.} \tag{3.5}$$

Solutions of these equations have a number of interesting features. For flow past a sphere, for example, Oseen pointed out that, unlike solutions of the Stokes equations, the solution of (3.2–5) is not symmetrical about a plane normal to the velocity at infinity. In fact, the solution displays a "wake" behavior: outside of a paraboloidal region around and downstream from the sphere, the flow is essentially that of an ideal fluid. Outside this wake, the flow becomes purely radial near infinity, and this is compensated for by a kind of "suction" behind the sphere, where the flow tends to follow the sphere.

The details of the Oseen flow past a sphere are complicated, as the above remarks perhaps make clear. To my knowledge, no one has solved the Oseen equation for flow past a sphere in closed form. However, a solution can be proved to exist, and approximations to it can be computed without too much difficulty. The details of one such approximation can be found in Lamb's book[1].

The solution in two dimensions is no simpler. Again, only approximate solutions are known. But the fact that even these approximations could be found led to the feeling that Oseen flow is not subject to the Stokes paradox. That is the case, but it is not so easy to prove[2].

This feature of Oseen flow is desirable, of course, and indicates that at least in some ways the conclusions one derives from Oseen flow are in accord with our intuition. However, Oseen flow is also subject to paradoxes. During the time that I first gave these lectures, Olmstead and Gautesen[3] proved the following result. Let $F(U)$ be the component of the force on a body in steady Oseen flow in the direction of the velocity U at infinity. Thus, $F(U)$ is the drag on the body. Then, Olmstead and Gautesen proved

$$F(U) = -F(-U).$$

In other words, whether one blows air from the front or the back of a body in Oseen flow, the drag is the same. Of course, in proving this result, no hypotheses are made with regard to symmetry of the body; if the body had appropriate symmetry properties, the result would be trivial.

It is, I think, the very complication of Oseen flow that caused this important paradox to lie undiscovered for so long. Since the flow is so complicated, hardly anyone computed any solutions of the Oseen equations for anything other than spheres, cylinders, and plates, whose symmetry is evident. Had anyone computed the drag due to Oseen flow for any unsymmetrical body, he could have conjectured immediately that the Olmstead–Gautesen paradox was correct.

4 Boundary layer flow

It is an experimental fact that low speed flows past most bodies resemble ideal fluid flows, at least outside of any wake that may be present. That is why the flows of Figures 8 through 10 appear so plausible. However, near a fixed wall, no flow can be ideal since both components of velocity have to vanish near a wall (see § 7.3). It is plausible to expect then that outside of a very thin layer around a body and outside of any wake that may exist, any flow is very

---

[1] *Op cit.*, § 342.

[2] For a proof, see R. Finn and D. Smith, On the linearized hydrodynamical equations in two dimensions. Archive for rational mechanics and analysis, 25 (1967) 1–25.

[3] A new paradox in viscous hydrodynamics. Archive for rational mechanics and analysis 29 (1968) 58–65. Olmstead has since sharpened the paradox still further. See: Force relationships and integral representations for the viscous hydrodynamical equations. Archive for rational mechanics and analysis, 31 (1968) 380–389.

nearly ideal. The layer around the body where the fluid is not approximately ideal is called the *boundary layer*. It goes without saying that this layer does not have a sharp boundary but that it merges slowly into the fluid as one moves away from the body.

For steady flow, the exact Navier–Stokes equations have the form

$$-\nu\nabla^2 s + (s\cdot\nabla)\,s = -\frac{1}{\varrho}\,\nabla p. \tag{4.1}$$

The boundary layer behavior is a manifestation of the fact that the kinematic viscosity is "small". Naturally, one has to define what ν is to be compared to call it small. Since ν clearly has dimensions (distance)^2/(time), it is usual to let U be a "characteristic" speed of the flow (say, the speed at infinity) and to let l be a "characteristic" length (for a sphere, its radius, for example) and to define a dimensionless parameter R by

$$R = \frac{Ul}{\nu}.$$

R is called the *Reynolds' number* of the flow, and ν can be called "small" if[1] $R \gg 1$.

For mathematical purposes, we can forget about Reynolds' number by fixing the speed at infinity and fixing the body in the flow, and looking on the solution of (4.1) as a function of the parameter ν appearing there. What presumably happens as $\nu \to 0$ is that away from the body and out of the wake, the solution of (4.1) tends to the solution of the equation obtained by setting $\nu = 0$. However, this can *not* be the case in the boundary layer. Thus, in the neighborhood of a fixed wall, say, s depends in non-uniform way on ν.

Consider a two-dimensional flow past the semi-infinite flat wall $y = 0$, $x > 0$. We take the velocity at (minus) infinity to be parallel to the x-axis and equal in magnitude to U.

We wish the fact that the flow is (or should be) nearly uniform outside the boundary layer to be manifested in our notation. Since everything interesting that happens does so when y is small, we stretch the y-variable while leaving the x-variable alone. Thus we write

$$x = \xi$$

$$y = \nu^\alpha\,\eta,$$

where α is a positive parameter.

---

[1] To give some idea of the magnitudes involved, for a one foot sphere moving one mile an hour in water, R is over 130,000. For the same sphere moving in air, R is about 9000.

In these coordinates, (4.1) becomes

$$-vu_{\xi\xi} - v^{1-2\alpha}\, u_{\eta\eta} + uu_{\xi} + v^{-\alpha}\, vu_{\eta} = -\frac{1}{\varrho}\, p_{\xi}$$

$$-vv_{\xi\xi} - v^{1-2\alpha}\, v_{\eta\eta} + uv_{\xi} + v^{-\alpha}\, vv_{\eta} = -\frac{v^{-\alpha}}{\varrho}\, p_{\eta}, \tag{4.2}$$

and the continuity equation becomes

$$v^{\alpha}\, u_{\xi} + v_{\eta} = 0. \tag{4.3}$$

We expect the motion to be mainly horizontal—that is, we expect v to be small. In accord with this idea, we write

$$v = v^{\beta}\, w,$$

for some $\beta > 0$. β can be evaluated by means of (4.3). In this equation, we wish the fact that a term is small to be expressed by the occurrence of a power of v in that term. The functions u and w that appear are themselves not to be especially small. Therefore, unless we wish one of the terms in (4.3) to dominate the other, we must set $\beta = \alpha$. (4.2) and (4.3) then become

$$-vu_{\xi\xi} - v^{1-2\alpha}u_{\eta\eta} + uu_{\xi} + wu_{\eta} = -\frac{1}{\varrho}\, p_{\xi} \tag{4.4}$$

$$-v^{1+2\alpha}\, w_{\xi\xi} - vw_{\eta\eta} + v^{2\alpha}\, uw_{\xi} + v^{2\alpha}\, ww_{\eta} = -\frac{1}{\varrho}\, p_{\eta} \tag{4.5}$$

$$u_{\xi} + w_{\eta} = 0. \tag{4.6}$$

Consider (4.5). If w, u, and their derivatives are bounded as $v \to 0$ (which we assume), then we must have $p_{\eta} \to 0$ as $v \to 0$. As a first approximation, we set

$$p_{\eta} = 0. \tag{4.7}$$

Now, outside of the boundary layer, we want the flow to approximate steady flow, in which the pressure is constant. This means that as $\eta \to \infty$, p must approach a constant. However, (4.5) says that p is independent of η. Therefore, we must set p equal everywhere to a constant, so that (4.4) becomes

$$-vu_{\xi\xi} - v^{1-2\alpha}\, u_{\eta\eta} + uu_{\xi} + wu_{\eta} = 0. \tag{4.8}$$

α is still free. If it is chosen such that $1 - 2\alpha > 0$, the first two terms in (4.8) are small, and we must solve

$$uu_{\xi} + wu_{\eta} = 0$$

approximately, along with (4.6). The only solution of these equations with zero velocity on the plate is the trivial one. Therefore, we must assume $1 - 2\alpha \leqslant 0$.

Suppose we choose $1 - 2\alpha < 0$. Then, the dominant term in (4.8) is the second, and we have

$$u_{\eta\eta} = 0,$$

approximately. If u is to be zero when $\eta = 0$, then, we must have

$$u = c\eta$$

where c is a constant. If u is to be bounded as $\eta \to \infty$, however, c must be zero. Therefore, again we obtain only a trivial answer.

Thus, we must choose $\alpha = \frac{1}{2}$, and the equations we must solve are

$$\left.\begin{aligned}
u_{\eta\eta} &= uu_\xi + wu_\eta, \\
u_\xi + w_\eta &= 0.
\end{aligned}\right\} \tag{4.9}$$

Equations (4.9) are Prandtl's *boundary layer equations*. I have derived them for the semi-infinite plate; naturally they must be modified slightly for other configurations.

The preceding derivation of the boundary layer equations is as plausible as I can make it. But even so, I have had to give the game away in a number of places. The quantity $1 - 2\alpha$ cannot be negative, for instance, "if u is to be bounded as $\eta \to \infty$". Although we can have no doubt that the physical phenomenon we are presumably discussing has this property, we have no *a priori* guarantee that the solutions of the Navier–Stokes equations possess it. That is something that must be proved. Moreover, it is clear that other arguments are possible, arguments that lead to other equations. It is only when we know the result we want that we can make an argument of the sort we have used.

Lamb[1] expressed very well the exasperation one must feel with these arguments by the faint praise he can give them. After giving a very brief argument leading to (4.9), he says, "The approximations are explained in greater detail by Blasius; they may be justified in the last resort by comparison with the results deduced." The mathematician's exasperation aside, though, there is no doubt that "the results deduced" from the boundary layer equations are in close accord with experiment[2].

There have been numerous attempts to justify the boundary layer equations, none of them, I think, completely successful, although scattered results are known[3]. The justification of the boundary layer equations remains a most important problem for the mathematician.

---

[1] *Op. cit.*, § 371 a.

[2] To see the kind of detailed physical conclusions one can draw using the boundary layer equations, see Hermann Schlichting, Boundary-layer theory. McGraw-Hill, 1968.

[3] The most recent results in this direction are due, I think, to Paul Fife, The generation of a boundary layer in hydrodynamics, and also Considerations regarding the mathematical basis for Prandtl's boundary layer theory. Both are in Archive for Rational Mechanics and Analysis, 21 (1966) 286–302, and 28 (1968) 184–216.

PART II

A TASTE OF THE MODERN THEORY

Introduction

THIS ENTIRE book is devoted to a study of the single set of partial differential equations (7.2.3–2), first when they describe ideal and then viscous fluids. Almost all the material covered to this point was discovered before the first World War, much of it in the eighteenth and nineteenth centuries. In this, Part II we, consider some of the more recent work on the Navier–Stokes equations.

The problems studied in a given age of science reflect, to a great extent, the techniques available. This truism works two ways. First, of course, there is the fact that a study begun before its time will often leave little direct trace, although it may influence the development of the appropriate techniques and thereby the eventual solution of the problem considered. More important is the fact that successful techniques, by the mere fact of their success, generate problems solvable by their use and attract people to study them. There are few clearer examples of this than the techniques of modern functional analysis. Much of the modern work on fluid mechanics is devoted to clarifying some of the many paradoxes associated with the theory. Much of it addresses itself to mere existence and uniqueness questions. In large measure, I think this is the case because functional analysis is eminently suited to answering such questions.

That the questions considered are determined by the techniques available is not a reason for despising them. Mathematical work can be of tremendous importance in the physical conceptions one has about fluids. Without the d'Alembert (and other) paradoxes, who would have thought it necessary to study more intricate models than the ideal fluid? However, it is usually through *paradoxes* that mathematical work has the greatest influence on physics. In terms of existence and uniqueness theory, this means that the most important thing to discover is what is *not* true. When one proves the Navier–Stokes equations have solutions, the physicist yawns. If one can prove these solutions are not unique (say), he opens his eyes instead of his mouth. Thus, when we prove existence theorems, we are only telling the world where paradoxes are not and perhaps sweeping away some of the mist that surrounds the area where they are.

What I am calling the modern theory began with the work of Leray[1], who

---

[1] J. Leray, Etude de diverses équations intégrales non linéaires et de quelques problèmes que pose l'hydrodynamique. J. Math. Pures Appl. 12 (1933) 1–82. Essai sur les mouvements plans d'un liquide visqueux que limitent des parois, J. Math. Pures Appl. 13 (1934) 331–418. Sur le mouvement d'un liquide visqueux emplissant l'espace. Acta Math. 63 (1934) 193–248.

put the Navier–Stokes equations in the context of functional analysis. Since Leray's papers appeared, many authors have taken the setting he supplied and within it have solved a number of different problems in various ways. In Part II, we consider a small section of this work. In chapter 10, we present the nomenclature and notation that will be used in the rest of the book, as well as a number of lemmas, mostly in the form of inequalities, that prove generally useful. Then, in chapter eleven, we begin in earnest, proving the existence of E. Hopf's weak solution. Then, we derive a criterion, due to Serrin, for the uniqueness of Hopf's solution. In the next chapter, we consider the question of where the weak solution is actually a solution and discuss, a little, where points of non-uniqueness might appear.

At the end, in chapter fourteen, we show that functional analytic techniques can be used for more than just existence and uniqueness, but also to derive certain qualitative properties of viscous fluid flow.

CHAPTER 10

Preliminaries

1 To start

IN THE NEXT chapters, we need a great deal of machinery, primarily in the form of inequalities between elements of various function spaces. Most of this is in the toolbox of every analyst, but some of it may not be at the command of any particular one. Therefore, we lay out most of these tools here. Because that is the purpose of this chapter, much of it appears unmotivated on first reading. Therefore, it is well simply to skim the chapter the first time around, paying attention mostly to the definitions and notation introduced; the results can be returned to when needed in later chapters. As an aid to this process, a table summarizing most of the notation introduced is provided at the end of the chapter.

Before going on, we state that the lower case letter c, with or without subscripts, is reserved as a generic name for a positive constant. If the letter c appears in a formula, it means that the formula is correct for some positive constant c. The same letter c is often used to denote different constants. This convention is necessary because of the large number of constants, whose exact numerical value is irrelevant, arising in the sequel.

In the rest of the book, V denotes a domain in R^3, usually bounded, unless something explicit to the contrary is stated. V is to be thought of as the domain occupied by the fluid, as before, the closure and the boundary of V are denoted by $\bar{V}$ and ∂V, respectively. In some of the results further on, it is assumed that ∂V is smooth. We are never explicit about how smooth ∂V need be, and different degrees of smoothness suffice for the different results, but in all cases ∂V having three continuous derivatives is more than enough.

2 Spaces of functions

The set of all functions with values in R^3 having n continuous derivatives in a set Σ is denoted by $C^n(\Sigma)$. Σ is usually V, $\bar{V}$, or R^3. If the dependence on Σ is suppressed in the notation, it is to be understood that Σ is V. Thus, for example, $C^n = C^n(V)$. Of course, $C^0(\Sigma)$ is just the set of continuous functions on Σ. Given any set of functions, the subscript zero is used to denote those functions in the set having compact support. Thus, C_0^n is just the set of functions defined in V, having n continuous derivatives, and zero in a neighborhood of ∂V (and of infinity if V is unbounded). We make the space C_0^0 a normed linear space by defining the norm

$$\max_{q \in V} |\varphi(q)|$$

on it. Here, $|\varphi(q)|$ denotes the Euclidean length of the element $\varphi(q)$ of C_0^0.

Given a set of functions denoted by a certain letter, the same letter in script is used to denote the functions in the set that are divergence-free. Thus, if $1 \leqq n \leqq \infty$,

$$\mathscr{C}^n = \{\varphi \in C^n(V): \nabla \cdot \varphi = 0\},$$

$$\mathscr{C}_0^n = \{\varphi \in C_0^n(V): \nabla \cdot \varphi = 0\}.$$

In all that follows, the letter p, with or without subscripts, is a real number greater than or equal to one. If $p \geqslant 1$ and $\varphi \in C_0^\infty$, we write

$$\|\varphi\|_p = \left(\int_V |\varphi|^p \, dq\right)^{1/p}.$$

If $1 \leqq p < \infty$, we denote by L^p the completion of the space C_0^∞ with respect to the norm $\|\varphi\|_p$. When $p = \infty$, we make the usual modification: L^∞ is the set of all measurable, essentially bounded functions in V with the norm

$$\|\varphi\|_\infty = \operatorname*{ess\,sup}_{q \in V} |\varphi(q)|.$$

If $1 \leqslant p \leqslant \infty$, we denote the index dual to p by p', so that

$$\frac{1}{p} + \frac{1}{p'} = 1.$$

The space L^2 is, of course, a Hilbert space. We denote the scalar product in L^2 by parentheses:

$$(\varphi, \psi) = \int_V \varphi \cdot \psi \, dq.$$

As in the earlier paragraph, the set of divergence-free functions in L^2 is denoted by a script letter. Precisely, $\mathscr{L}^2$ is the closure of $\mathscr{C}_0^\infty$ in L^2. $\mathscr{L}^2$ is a linear subspace of L^2. Another subspace we need is the set of all gradients. We denote this set by C:

$$G = \{\varphi \in C^\infty \cap L^2: \varphi = \nabla \pi \text{ for some real-valued function } \pi\}.$$

There is an important relation between the two subspaces $\mathscr{L}^2$ and G. We state this relation in the following lemma.

LEMMA 2.1 *Let $G^\perp$ be the orthocomplement of G in the Hilbert space L^2. Then, $\mathscr{L}^2 = G^\perp$.*

Proof   Let $\varphi \in \mathscr{C}_0^\infty$. Then, $\nabla \cdot \varphi = 0$, so that whenever $\nabla \pi \in G$,

$$\int_V \varphi \cdot \nabla \pi \, dq = \int_V \nabla \cdot (\varphi \pi) \, dq = 0,$$

by the divergence theorem. This shows that $\mathscr{C}_0^\infty \subset G^\perp$; closing in L^2, we find $\mathscr{L}^2 \subset G^\perp$.

For the opposite inclusion, let $\varphi \in (\mathscr{L}^2)^\perp$. Then, whenever $\psi \in \mathscr{C}_0^\infty$,

$$\int_V \varphi \cdot \psi \, dq = 0.$$

We define φ to be zero outside V, obtaining thereby a function in $(\mathscr{L}^2(R^3))^\perp$. If $\psi \in \mathscr{C}_0^\infty(R^3)$,

$$\int_{R^3} \varphi \cdot \psi \, dq = 0. \tag{2.1}$$

Let $\chi \in C_0^\infty(R^3)$. Then, $\psi = \nabla \times \chi \in \mathscr{C}_0^\infty(R^3)$. Let q and q_0 be points in R^3, ∇_q the gradient with respect to the variable q, and ∇_{q_0} the gradient with respect to the variable q_0. (2.1) gives

$$0 = \int_{R^3} \varphi(q) \cdot [\nabla_q \times \chi \, (q_0 - q)] \, dq$$

$$= - \int_{R^3} \varphi(q) \cdot [\nabla_{q_0} \times \chi \, (q_0 - q)] \, dq$$

$$= \nabla_{q_0} \cdot \int_{R^3} \varphi(q) \times \chi \, (q_0 - q) \, dq.$$

This shows that the integral appearing in the last step is divergence free. Since it is clearly in $C_0^\infty(R^3)$ if χ is, the integral is actually in $\mathscr{C}_0^\infty(R^3)$. By the part of the lemma already proved, then, $\nabla \pi \in G$ implies

$$0 = \int_{R^3} \nabla_{q_0} \pi \, (q_0) \cdot \int_{R^3} \varphi(q) \times \chi \, (q_0 - q) \, dq \, dq_0$$

$$= \int_{R^3} \nabla_{q_0} \pi \, (q_0) \cdot \int_{R^3} \varphi \, (q_0 - q) \times \chi(q) \, dq \, dq_0$$

$$= \int_{R^3} \chi(q) \cdot \int_{R^3} [\nabla_{q_0} \pi \, (q_0)] \times \varphi \, (q_0 - q) \, dq_0 \, dq$$

$$= \int_{R^3} \chi(q) \cdot \int_{R^3} [\nabla_{q_0} \pi \, (q + q_0)] \times \varphi(q_0) \, dq_0 \, dq.$$

χ is an arbitrary smooth function with compact support. Therefore,

$$\int_{R^3} [\nabla_{q_0}\pi\,(q + q_0)] \times \varphi(q_0)\,dq_0 = 0.$$

Since π is smooth, we may set $q = 0$ in this formula. Then, writing q for q_0, we find

$$\int_{R^3} \nabla\pi \times \varphi\,dq = 0.$$

If, in a given coordinate system, $\varphi = (\varphi^1, \varphi^2, \varphi^3)$, define (ψ^1, ψ^2, ψ^3) by

$$\psi^1_x = \varphi^1, \quad \psi^2_y = \varphi^2, \quad \psi^3_z = \varphi^3.$$

Substituting this in the last displayed equation and integrating by parts, we find that if π has compact support,

$$\int_{R^3} (\psi^3 - \psi^2)\,\pi_{yz}\,dq = 0,$$

$$\int_{R^3} (\psi^1 - \psi^3)\,\pi_{xz}\,dq = 0,$$

$$\int_{R^3} (\psi^2 - \psi^1)\,\pi_{xy}\,dq = 0.$$

Since π is otherwise arbitrary, this shows that $\psi^1 = \psi^2 = \psi^3$. Write ψ for the common function ψ^1, ψ^2, ψ^3. The definition of these functions shows that

$$\varphi^1 = \psi_x, \quad \varphi^2 = \psi_y, \quad \varphi^3 = \psi_z.$$

Thus, φ is a gradient. This shows that $(\mathscr{L}^2)^{\perp} \subset \bar{G}$ and completes the proof of lemma 2.1.

If φ is a function with values in R^3, the 3×3 matrix of all partial derivatives of the components of φ is denoted by $\nabla\varphi$, just as in § 3.7. If $A = (a_{ij})$ and $B = (b_{ij})$ are matrices, we define

$$A \cdot B = \sum_{i,j=1}^{3} a_{ij}b_{ij}.$$

$$|A|^2 = A \cdot A.$$

Notice that $A \cdot B$ is *not* ordinary matrix multiplication, but is the sum of the products of corresponding entries in A and in B. For $\varphi \in C_0^\infty$, we define

$$\|\nabla\varphi\|_p = \left(\int_V |\nabla\varphi|^p\,dq \right)^{1/p}.$$

The completions of C_0^∞ and $\mathscr{C}_0^\infty$ in this norm[2] are denoted by $H_0^{1,p}$ and $\mathscr{H}_0^{1,p}$, respectively. $H_0^{1,2}$ is clearly a Hilbert space. If φ and ψ are in $H_0^{1,2}$, we denote the scalar product of φ and ψ in $H_0^{1,2}$ by

$$(\nabla\varphi, \nabla\psi) = \int_V \nabla\varphi \cdot \nabla\psi \, dq.$$

In all that follows, the Hölder inequality is used over and over again. As a reminder, we state it here; the proof can be found in any text on the theory of functions of a real variable.

LEMMA 2.2   *If $\varphi \in L^p$ and $\psi \in L^{p'}$, then*

$$\left| \int_V \varphi \cdot \psi \, dq \right| \leqslant \|\varphi\|_p \|\psi\|_{p'};$$

here, q can be a variable in R^n for any n.

An immediate consequence is

LEMMA 2.3   *If V is bounded and $p_1 \geqslant p_2$, then $L^{p_2} \subset L^{p_1}$, and*

$$\|\varphi\|_{p_1} \leqslant c \, \|\varphi\|_{p_2}. \tag{2.2}$$

Proof   That $L^{p_2} \subset L^{p_1}$ follows from (2.2). To prove (2.2), consider

$$\|\varphi\|_{p_1} = \left(\int_V |\varphi|^{p_1} \, dq \right)^{1/p_1}$$

$$\leqslant \left(\int_V dq \right)^{1/p_1 - 1/p_2} \left(\int_V |\varphi|^{p_2} \, dq \right)^{1/p_2}$$

$$= c \, \|\varphi\|_{p_2}.$$

Hölder's inequality is used in the second step.

LEMMA 2.4   *Let $\varphi \in L^{p_1} \cap L^{p_3}$, where $p_1 < p_3$. Then, $\varphi \in L^{p_2}$ whenever $p_1 \leqslant p_2 \leqslant p_3$, and*

$$\|\varphi\|_{p_2}^{1/p_1 - 1/p_3} \leqslant \|\varphi\|_{p_1}^{1/p_2 - 1/p_3} \|\varphi\|_{p_3}^{1/p_1 - 1/p_2}. \tag{2.3}$$

Proof   If $0 \leqslant p \leqslant p_2$, we have

$$\|\varphi\|_{p_2} = \left(\int_V |\varphi|^p |\varphi|^{p_2 - p} \, dq \right)^{1/p_2} \leqslant \|\varphi\|_{pp_4}^{p/p_2} \|\varphi\|_{(p_2 - p)p_4'}^{1 - p/p_2},$$

by lemma 2.2. Setting $pp_4 = p_1$, $(p_2 - p) p_4' = p_3$, the lemma follows.

---

[2] $\|\nabla\varphi\|_p$ is a norm because φ is zero on ∂V.

LEMMA 2.5 *If $a \geq 0$, $b \geq 0$, $\eta > 0$, and $p > 1$, then*

$$ab \leq \frac{a^p}{p\eta} + \frac{\eta^{p'} b^{p'}}{p'}. \tag{2.4}$$

Proof We need only prove (2.4) for $\eta = 1$, since (2.4) for any $\eta > 0$ follows from (2.4) with $\eta = 1$ by replacing a by a/η and b by ηb. Now, it is easy to prove that for $x \geq 0$, the function $(x/p) + (x^{1-p'}/p')$ has its only minimum at $x = 1$. Therefore,

$$\frac{x}{p} + \frac{x^{1-p'}}{p'} \geq \frac{1}{p} + \frac{1}{p'} = 1. \tag{2.5}$$

If $b = 0$, (2.4) is trivial. If $b \neq 0$, set $x = a^{p-1}/b$ in (2.5). Upon simplification, the result is (2.4) with $\eta = 1$.

LEMMA 2.6 *Let $\varphi \in H_0^{1,p}$, where $1 \leq p < 3$. Then,*

$$\|\varphi\|_{3p/(3-p)} \leq c \|\nabla\varphi\|_p. \tag{2.6}$$

Therefore, $H_0^{1,p} \subset L^{3p/(3-p)}$, with continuous injection.
(2.6) is known as *the Sobolev inequality.*
To prove lemma 2.6, let $\psi \in C_0^1(R^3)$. Then,

$$\psi = \int_{-\infty}^{x} \psi_x \, dx,$$

so that

$$|\psi| \leq \int_{-\infty}^{\infty} |\psi_x| \, dx.$$

The same formula remains correct when x is replaced by y and by z. Therefore,

$$|\psi|^{3/2} \leq \left(\int_{-\infty}^{\infty} |\psi_x| \, dx\right)^{1/2} \left(\int_{-\infty}^{\infty} |\psi_y| \, dy\right)^{1/2} \left(\int_{-\infty}^{\infty} |\psi_z| \, dz\right)^{1/2}.$$

Integrating this inequality with respect to x and using the Schwarz inequality, we find

$$\int_{-\infty}^{\infty} |\psi|^{3/2} \, dx$$

$$\leq \left(\int_{-\infty}^{\infty} |\psi_x| \, dx\right)^{1/2} \left(\int_{-\infty}^{\infty}\int_{-\infty}^{\infty} |\psi_y| \, dy\,dx\right)^{1/2} \left(\int_{-\infty}^{\infty}\int_{-\infty}^{\infty} |\psi_z| \, dz\,dx\right)^{1/2}.$$

Repeating this process, but integrating this time with respect to y, we obtain

$$\int_{-\infty}^{\infty} \int_{-\infty}^{\infty} |\psi|^{3/2} \, dx \, dy$$

$$\leqslant \left(\int_{-\infty}^{\infty} \int_{-\infty}^{\infty} |\psi_y| \, dy \, dx \right)^{1/2} \left(\int_{-\infty}^{\infty} \int_{-\infty}^{\infty} |\psi_x| \, dx \, dy \right)^{1/2} \left(\int_{R^3} |\psi_z| \, dq \right)^{1/2}.$$

Integrate for the last time with respect to z. The result is

$$\int_{R^3} |\psi|^{3/2} \, dq \leqslant \left(\int_{R^3} |\psi_x| \, dq \right)^{1/2} \left(\int_{R^3} |\psi_y| \, dq \right)^{1/2} \left(\int_{R^3} |\psi_z| \, dq \right)^{1/2}.$$

Now, use lemma 2.5 with $p = \frac{3}{2}$ to obtain

$$\int_{R^3} |\psi|^{3/2} \, dq \leqslant \frac{2}{3} \left(\int_{R^3} |\psi_x| \, dq \right)^{3/4} \left(\int_{R^3} |\psi_y| \, dq \right)^{3/4} + \frac{1}{3} \left(\int_{R^3} |\psi_z| \, dq \right)^{3/2}.$$

Applying lemma 2.5 (with $p = 2$) to the first term on the right, we find

$$\int_{R^3} |\psi|^{3/2} \, dq$$

$$\leqslant \frac{1}{3} \left(\int_{R^3} |\psi_x| \, dq \right)^{3/2} + \frac{1}{3} \left(\int_{R^3} |\psi_y| \, dq \right)^{3/2} + \frac{1}{3} \left(\int_{R^3} |\psi_z| \, dq \right)^{3/2}$$

$$\leqslant \left(\int_{R^3} |\nabla \psi| \, dq \right)^{3/2}. \tag{2.7}$$

It should be noted that (2.7) is just (2.6) when $V = R^3$ and $p = 1$.

Let $\varphi \in C_0^\infty(V)$. Since $p \geqslant 1$, $2p/(3 - p) \geqslant 1$, so that $|\varphi|^{2p/(3-p)} \in C_0^1(V)$. Define

$$\psi(q) = \begin{cases} |\varphi(q)|^{2p/(3-p)} & \text{if } q \in V, \\ 0 & \text{if } q \notin V. \end{cases}$$

Then, $\psi \in C_0^1(R^3)$. Substituting this function ψ in (2.7) and using Hölder's inequality, we obtain (2.6) for $\varphi \in C_0^\infty$. But this implies (2.6) for all $\varphi \in H_0^{1,p}$ since C_0^∞ is dense in $H_0^{1,p}$.

Lemmas 2.3 and 2.6 together give

LEMMA 2.7 *Let $\varphi \in H_0^{1,p}$, where $1 \leqslant p < 3$. If V is bounded, then*

$$\|\varphi\|_{p_1} \leqslant c \, \|\nabla \varphi\|_p$$

whenever

$$1 \leqslant p_1 \leqslant \frac{3p}{3 - p}.$$

We also have

LEMMA 2.8   *Let $\varphi \in H_0^{1,p}$, where $p > 3$. Then*

$$|\varphi(q_2) - \varphi(q_1)| \leqslant c \, |q_2 - q_1|^{1-p/3} \, \|\nabla\varphi\|_p. \tag{2.8}$$

Therefore, every element in $H_0^{1,p}$ is continuous in $\bar{V}$.

Proof   First, assume $\varphi \in C_0^\infty$. As usual, let $B_r(q)$ and $S_r(q)$ denote the ball and the sphere in R^3 of radius r and center q. If q_1 and q_2 are two fixed points in R^3, we write $B_1 = B_{|q_2 - q_1|}(q_1)$, $B_2 = B_{|q_2 - q_1|}(q_2)$, and $\Sigma = B_1 \cap B_2$. Let $|\Sigma|$ denote the volume of Σ. Extend φ to a function in $C_0^\infty(R^3)$ by defining it as zero outside V. We have

$$|\varphi(q_2) - \varphi(q_1)| \cdot |\Sigma| = \int_\Sigma |\varphi(q_2) - \varphi(q_1)| \, dq$$

$$\leqq \int_\Sigma |\varphi(q_2) - \varphi(q)| \, dq$$

$$+ \int_\Sigma |\varphi(q) - \varphi(q_1)| \, dq. \tag{2.9}$$

Introduce polar coordinates with the origin at q_2 by writing $q = q_2 + rS$, where $r = |q - q_2|$ and S is a unit vector. The first integral on the right of (2.9) is bounded by

$$\int_{B_2} \left(\int_0^r |\varphi_r(q_2 + r_1 S)| \, dr_1 \right) dq$$

$$= \int_{|S|=1} \int_0^{|q_2-q_1|} r^2 \int_0^r |\varphi_r(q_2 + r_1 S)| \, dr_1 \, dr \, dS$$

$$= \int_{|S|=1} \int_0^{|q_2-q_1|} \frac{|q_2 - q_1|^3 - r_1^3}{3} |\varphi_r(q_2 + r_1 S)| \, dr_1 \, dS$$

$$\leqslant \frac{1}{3} |q_2 - q_1|^3 \int_{B_2} |\varphi_r(q)| \frac{dq}{|q - q_2|^2}. \tag{2.10}$$

Now,

$$\int_{B_2} \frac{dq}{|q - q_2|^{2p'}} = \int_{|S|=1} \int_0^{|q_2-q_1|} r^{2(1-p')} \, dr \, dS$$

$$= c \, |q_2 - q_1|^{3 - 2p'},$$

since $p > 3$. Therefore, Hölder's inequality applied to (2.10) shows that

$$\int_\Sigma |\varphi(q_2) - \varphi(q)| \, dq \leq c \, |q_2 - q_1|^{1 + 3/p'} \left[\int_{B_2} |\varphi_r(q)|^p dq \right]^{1/p}$$

$$\leqq c \, |q_2 - q_1|^{1 + 3/p'} \|\nabla\varphi\|_p.$$

A similar estimate can be found for the second integral on the right of (2.9). Since $|\Sigma|$ is clearly proportional to $|q_2 - q_1|^3$, we find

$$|\varphi(q_2) - \varphi(q_1)| \leq c\,|q_2 - q_1|^{3/p'-2}\,\|\nabla\varphi\|_p$$
$$= c\,|q_2 - q_1|^{1-3/p}\,\|\nabla\varphi\|_p.$$

This is (2.8) for $\varphi \in C_0^\infty$. The fact that C_0^∞ is dense in $H_0^{1,p}$ now implies (2.8) for all $\varphi \in H_0^{1,p}$, and (2.8) in turn implies the continuity of φ.

The next result is an inequality due to Friedrichs.

LEMMA 2.9   *Let V be bounded. Then, there exists a sequence $\{\omega_n\} \subset C^\infty(\bar{V})$ with the following property. Given any $\varepsilon > 0$, there exists an $N > 0$ such that*

$$\|\varphi\|_2^2 \leq \sum_1^N |(\varphi, \omega_n)|^2 + \varepsilon\,\|\nabla\varphi\|_2^2$$

for all $\varphi \in H_0^{1,2}$.

Proof   As usual, it suffices to prove the lemma for $\varphi \in C_0^\infty$ since C_0^∞ is dense in $H_0^{1,2}$. Since V is bounded, there is a cube C: $-a < x < a$, $-a < y < a$, $-a < z < a$ such that $V \subset C$. If $\varphi \in C_0^\infty$, we define φ to be zero outside V to obtain a function $\varphi \in C_0^\infty(C)$.

The sequence

$$\left\{ \frac{1}{a^3} \left(\cos \frac{j\pi x}{2a} \right) \left(\cos \frac{k\pi y}{2a} \right) \left(\cos \frac{l\pi z}{2a} \right) \right\},$$

where j, k, and l are integers, is complete in $L^2(C)$. Enumerate this sequence to obtain a sequence $\{\omega_n(q)\}$. The enumeration can obviously be made in such a way that any of the indefinite integrals

$$\int \omega_n\, dx, \int \omega_n\, dy, \int \omega_n\, dz$$

is less than ε_n, where ε_n is a sequence of real numbers decreasing to zero as $n \to \infty$.

$\{\omega_n\}$ is orthonormal and complete in $L^2(C)$. Therefore, if $\varphi \subset C_0^\infty(C)$,

$$\varphi = \sum_1^\infty (\varphi, \omega_n)\,\omega_n.$$

Now, integrating by parts,

$$(\varphi, \omega_n) = (\varphi', \bar{\omega}_n),$$

where φ' is one of the derivatives φ_x, φ_y, φ_z, while $\bar{\omega}_n$ is an indefinite integral of ω_n. Since each ω_n is a product of cosines, $\bar{\omega}_n = \varepsilon_n \theta_n$, where $\{\theta_n\}$ is also an orthonormal sequence in $L^2(C)$. Therefore, we have,

$$\|\varphi\|_2^2 = \sum_1^N |(\varphi, \omega_n)|^2 + \sum_{N+1}^\infty \varepsilon_n |(\varphi', \theta_n)|^2$$

$$\leq \sum_1^N |(\varphi, \omega_n)|^2 + \varepsilon_{N+1} \|\varphi'\|_2^2$$

by Bessel's inequality. Since φ' is one of the derivatives of φ, while $\varepsilon_n \downarrow 0$, the lemma follows.

A well known result, valid for any Hilbert space is

LEMMA 2.10   *Let H be a Hilbert space with norm $||| \cdot |||$ and scalar rproduct $((\cdot, \cdot))$. If $\{\varphi_n\}$ is any sequence in H, converging weakly to φ, then*

$$|||\varphi||| \leq \lim \inf |||\varphi_n|||.$$

Proof   If $|||\varphi||| = 0$, the lemma is trivial. Suppose, then, that $|||\varphi||| \neq 0$. Since $\{\varphi_n\}$ converges weakly to φ, $\{((\varphi_n, \varphi))\}$ converges to $|||\varphi|||^2$. Therefore, given any $\varepsilon > 0$,

$$|||\varphi|||^2 \leq (\varphi_n, \varphi) + \varepsilon |||\varphi|||$$

$$\leq |||\varphi_n||| \cdot |||\varphi||| + \varepsilon |||\varphi|||,$$

if n is large enough. Thus,

$$|||\varphi||| \leq |||\varphi_n||| + \varepsilon$$

for n large enough. The lemma follows from this.

Let $\mathscr{C}_*^n$ be the set of all functions $\varphi \in \mathscr{C}^n(V) \cap C^0(\bar{V})$ such that $\varphi(q) = 0$ for $q \in \partial V$. Let P be orthogonal projection in L^2 onto the subspace $\mathscr{L}^2$. Then, the quantity $\|P\nabla^2\varphi\|_2$ is a norm on the space $\mathscr{C}_*^n$ if $n \geq 2$. To show this, we only have to prove that $\|P\nabla^2\varphi\|_2 = 0$ implies $\varphi = 0$. Let $\varphi \in \mathscr{C}_*^n$ and $\|P\nabla^2\varphi\|_2 = 0$. Then, since $\mathscr{C}_*^n \subset \mathscr{L}^2$, we have

$$0 = (P\nabla^2\varphi, \varphi) = (\nabla^2\varphi, P\varphi) = (\nabla^2\varphi, \varphi) = -\|\nabla\varphi\|_2^2,$$

by Green's theorem. Thus, $\nabla\varphi = 0$ in V, and φ is constant. Since $\varphi = 0$ on ∂V, we see that φ is identically zero. Thus, $\|P\nabla^2\varphi\|_2$ is a norm. The completion of $\mathscr{C}_*^\infty$ in this norm is denoted by $\mathscr{H}^2$.

The operator $P\nabla^2$ has a natural extension to a map from $\mathscr{H}^2$ to $\mathscr{L}^2$. If $\varphi \in \mathscr{H}^2$, there is a sequence $\{\varphi_n\} \subset \mathscr{C}_*^\infty$ converging to φ in $\mathscr{H}^2$. Since $\{\varphi_n\}$ converges in $\mathscr{H}^2$, it is Cauchy there, which is to say that $\{P\nabla^2\varphi_n\}$ is Cauchy in $\mathscr{L}^2$. We can define $P\nabla^2\varphi$ as the limit in $\mathscr{L}^2$ of the sequence $\{P\nabla^2\varphi_n\}$. It is clear that this definition is independent of the sequence $\{\varphi_n\}$ in $\mathscr{C}_*^\infty$ chosen to represent φ.

Exercises

2.1 Show that if $\varphi \in H_0^{1,2} \cap C^0(\overline{V})$, and ∂V is smooth enough, then $\varphi = 0$ on ∂V.

2.2 Let V be a domain in R^2 instead of R^3. Show that if $\varphi \in H_0^{1,p}$ with $1 \leqq p < 2$, then

$$\|\varphi\|_{2p/(p-2)} \leqq c \, \|\nabla\varphi\|_p.$$

This is the Sobolev inequality in two dimensions.

2.3 Show that if $p > 1$, the dual of the Banach space $\mathscr{H}_0^{1,p}$ is $\mathscr{H}_0^{1,p'}$ in the sense that given any bounded linear functional F on $\mathscr{H}_0^{1,p}$, there exists a $\psi \in \mathscr{H}_0^{1,p'}$ such that

$$F(\varphi) = (\nabla\varphi, \nabla\psi).$$

2.4 Show that if $\varphi \in C^1(\overline{V})$ and ∂V is smooth enough, then $\varphi \in \mathscr{L}^2$ if and only if

$$\nabla \cdot \varphi = 0 \quad \text{in} \quad V,$$

$$n \cdot \varphi = 0 \quad \text{on} \quad \partial V,$$

where n is the normal to ∂V.

3 Functions of time

Let B denote any of the spaces defined so far: C^n, L^p, $H_0^{1,p}$, etc. If B is one of the spaces for which a norm has been defined, we write $\|\cdot\|_B$ for the norm on B. We consider now functions $\varphi : t \to \varphi(t)$ defined for all t in an interval I of real numbers, with φ taking values in B. We say that φ is *continuous* if, for all $t_0 \in I$, $\|\varphi(t) - \varphi(t_0)\|_B \to 0$ as $t \to t_0$. We denote the space of all continuous functions φ defined on I with values in B by $C^0(I; B)$. If $\varphi \in C^0(I; B)$ and $\varphi(t) = 0$ for t outside of a compact subset of I, we write $\varphi \in C_0^0(I; B)$. It should be noted that if I is a closed, bounded interval, then $C^0(I; B) = C_0^0(I; B)$. Also, if I is a half-open interval, say, $I = [0, T)$, then a function in $C_0^0(I; B)$ is zero in a neighborhood of $t = T$ but need *not* be zero at $t = 0$.

If $\varphi \in C^0(I; B)$, and if there exists a function $\dot\varphi \in C^0(I; B)$ such that

$$\left\| \frac{\varphi(t + h) - \varphi(t)}{h} - \dot\varphi(t) \right\| \to 0 \quad \text{as} \quad h \to 0,$$

we say that $\varphi \in C^1(I; B)$. Of course, we call $\dot\varphi$ the *derivative* of φ. If $\varphi \in C_0^0(I; B)$ and $\dot\varphi$ exists, we write $\varphi \in C_0^1(I; B)$.

We can also define spaces $C_0^0(I; B)$ and $C_0^1(I; B)$ when B is $\mathscr{C}_0^\infty$, a space that has no norm defined on it. We write $\varphi \in C_0^i(I; \mathscr{C}_0^\infty)$ $(i = 0, 1)$ if $\varphi \in C_0^i(I; C_0^0)$ and $\varphi(t) \in \mathscr{C}_0^\infty$ for all $t \in I$.

When B is normed, the space $C_0^0(I; B)$ has a norm associated with it, namely,

$$\left(\int_I \|\varphi(t)\|_B^q \, dt \right)^{1/q}.$$

The completion of $C_0^0 (I; B)$ with respect to this norm is a space that we call $L^q (I; B)$. Again, we make the usual modification for $q = \infty$: if $\|\varphi(t)\|_B$ is measurable and essentially bounded, we say that $\varphi \in L^\infty (I; B)$. When $B = L^p$, we write $\|\varphi\|_{p,q}$ for the norm in $L^q (I; L^p)$. Thus,

$$\|\varphi\|_{p,q} = \left(\int_I \|\varphi(t)\|_p^q \, dt \right)^{1/q},$$

and, of course,

$$\|\varphi\|_{p,\infty} = \operatorname*{ess\,sup}_{t \in I} \|\varphi(t)\|_p.$$

Similarly, when $B = H_0^{1,p}$, we write $\|\nabla\varphi\|_{p,q}$ for the norm in $L^q (I; H_0^{1,p})$:

$$\|\nabla\varphi\|_{p,q} = \left(\int_I \|\nabla\varphi (t)\|_p^q \, dt \right)^{1/q}.$$

One more space, and we're done with this tedious list. Let $\varphi \in C_0^1 ([0, T); \mathscr{H}^2)$. Then

$$\left[\|\nabla\varphi (0)\|_2^2 + \int_0^T e^{2at} (\|\dot\varphi(t)\|_2^2 + \|P\nabla^2\varphi (t)\|_2^2) \, dt \right]^{1/2}$$

is a (Hilbert) norm for any real a. The completion of $C_0^1 ([0, T); \mathscr{H}^2)$ with respect to this norm is denoted by $\mathscr{K} (a, T)$. It is clear that if $T < \infty$, any two of the spaces $\mathscr{K} (a, T)$ and $\mathscr{K} (b, T)$ are the same; we introduce the parameter a for the case $T = \infty$.

In chapter twelve, we need a theorem on mollifiers. To prove it, notice that if B is a Banach space and $\varphi \in C_0 (I; B)$, the integral

$$\int_I \varphi(t) \, dt \tag{3.1}$$

can be defined in the obvious way. One begins by partitioning the interval I, looking at the Riemann sums associated with (3.1) and then showing that because $\varphi \in C_0^0 (I; B)$, the sums converge to an element[3] of B. The integral (3.1) can then be extended to all $\varphi \in L^1 (I; B)$ by continuity. With this definition of the integral, we have

LEMMA 3.1   Let $\varphi (t, \tau)$ be a function defined on R^2 with values in a Banach space B, satisfying

$$\int_{-\infty}^{\infty} \int_{-\infty}^{\infty} \|\varphi (t, \tau)\|_B^p \, d\tau \, dt < \infty.$$

---

[3] For the details of an alternate construction, see Nelson Dunford and Jacob T. Schwartz, Linear Operators, I. Interscience publishers, Inc. New York, 1958.

Then,

$$\left[\int_{-\infty}^{\infty}\left\|\int_{-\infty}^{\infty}\varphi\,(t,\tau)\,d\tau\right\|_{B}^{p}\,dt\right]^{1/p}\leqslant\int_{-\infty}^{\infty}\left[\int_{-\infty}^{\infty}\|\varphi\,(t,\tau)\|_{B}^{p}\,dt\right]^{1/p}\,d\tau.$$

Proof. We have

$$\int_{-\infty}^{\infty}\left\|\int_{-\infty}^{\infty}\varphi\,(t,\tau)\,d\tau\right\|_{B}^{p}\,dt=\int_{-\infty}^{\infty}\left\|\int_{-\infty}^{\infty}\varphi\,(t,\tau)\,d\tau\right\|_{B}^{p-1}\left\|\int_{-\infty}^{\infty}\varphi\,(t,\sigma)\,d\sigma\right\|_{B}\,dt$$

$$\leqslant\int_{-\infty}^{\infty}\left\|\int_{-\infty}^{\infty}\varphi\,(t,\tau)\,d\tau\right\|_{B}^{p-1}\int_{-\infty}^{\infty}\|\varphi\,(t,\sigma)\|_{B}\,d\sigma\,dt$$

$$=\int_{-\infty}^{\infty}\left[\int_{-\infty}^{\infty}\left\|\int_{-\infty}^{\infty}\varphi\,(t,\tau)\,d\tau\right\|_{B}^{p-1}\|\varphi\,(t,\sigma)\|_{B}\,dt\right]\,d\sigma$$

$$\leqslant\int_{-\infty}^{\infty}\left[\int_{-\infty}^{\infty}\left\|\int_{-\infty}^{\infty}\varphi\,(t,\tau)\,d\tau\right\|_{B}^{p}\,dt\right]^{(p-1)/p}\times$$

$$\times\left[\int_{-\infty}^{\infty}\|\varphi\,(t,\sigma)\|_{B}^{p}\,dt\right]^{1/p}\,d\sigma,$$

by lemma 2.2. Therefore,

$$\int_{-\infty}^{\infty}\left\|\int_{-\infty}^{\infty}\varphi\,(t,\tau)\,d\tau\right\|_{B}^{p}\,dt$$

$$\leqq\left[\int_{-\infty}^{\infty}\left\|\int_{-\infty}^{\infty}\varphi\,(t,\tau)\,d\tau\right\|_{B}^{p}\,dt\right]^{(p-1)/p}\int_{-\infty}^{\infty}\left[\int_{-\infty}^{\infty}\|\varphi\,(t,\sigma)\|_{B}^{p}\,dt\right]^{1/p}\,d\sigma.$$

The lemma follows immediately from this.

Let $k(t)$ be a non-negative, real-valued function in $C_0^1\,((-\infty,\infty);R^1)$ such that

$$\int_{-\infty}^{\infty}k(t)\,dt=1. \tag{3.2}$$

If $k(t)$ is any such function, the functions

$$k_\varepsilon(t)=\frac{1}{\varepsilon}\,k\left(\frac{t}{\varepsilon}\right) \tag{3.3}$$

are called *mollifiers*. If $k_\varepsilon(t)$ is a mollifier and $\varphi(t)\in L^q((-\infty,\infty);B)$, we define the *convolution* of k_ε and φ by the formula

$$(k_\varepsilon*\varphi)\,(t)=\int_{-\infty}^{\infty}k_\varepsilon\,(t-\tau)\,\varphi(\tau)\,d\tau. \tag{3.4}$$

LEMMA 3.2 *Let B be a Banach space, and let $k_\varepsilon(t)$ be a mollifier. Then,*

$$\int_{-\infty}^{\infty} \|\varphi(t) - (k_\varepsilon * \varphi)(t)\|_B^q \, dt \to 0 \quad as \quad \varepsilon \downarrow 0$$

for all $\varphi \in L^q((-\infty, \infty); B)$.

Proof The definitions (3.3) and (3.4) give

$$\int_{-\infty}^{\infty} \|(k_\varepsilon * \varphi)(t)\|_B^q \, dt = \int_{-\infty}^{\infty} \left\| \int_{-\infty}^{\infty} k(\tau) \, \varphi(t - \varepsilon\tau) \, d\tau \right\|_B^q \, dt.$$

Therefore, lemma 3.1 shows that

$$\left[\int_{-\infty}^{\infty} \|(k_\varepsilon * \varphi)(t)\|_B^q \, dt \right]^{1/q} \leq \int_{-\infty}^{\infty} k(\tau) \left[\int_{-\infty}^{\infty} \|\varphi(t - \varepsilon\tau)\|_B^q \, dt \right]^{1/q} d\tau$$

$$= \left[\int_{-\infty}^{\infty} \|\varphi(t)\|_B^q \, dt \right]^{1/q} \tag{3.5}$$

by (3.2).

Since $C_0^0((-\infty, \infty); B)$ is dense in $L^q((-\infty, \infty); B)$, there is a function $\varphi_0 \in C_0^0((-\infty, \infty); B)$ such that

$$\left[\int_{-\infty}^{\infty} \|\varphi(t) - \varphi_0(t)\|_B^q \, dt \right]^{1/q} < \frac{\delta}{3} \tag{3.6}$$

for any preassigned $\delta > 0$. Also, (3.5) gives

$$\left[\int_{-\infty}^{\infty} \|(k_\varepsilon * \varphi_0)(t) - (k_\varepsilon * \varphi)(t)\|_B^q \, dt \right]^{1/q}$$

$$= \left[\int_{-\infty}^{\infty} \|(k_\varepsilon * \{\varphi_0 - \varphi\})(t)\|_B^q \, dt \right]^{1/q}$$

$$\leq \left[\int_{-\infty}^{\infty} \|\varphi_0(t) - \varphi(t)\|_B^q \, dt \right]^{1/q} < \frac{\delta}{3}. \tag{3.7}$$

Now, consider

$$\left[\int_{-\infty}^{\infty} \|\varphi_0(t) - (k_\varepsilon * \varphi_0)(t)\|_B^q \, dt \right]^{1/q}$$

$$= \left[\int_{-\infty}^{\infty} \left\| \int_{-\infty}^{\infty} k(\tau) [\varphi_0(t) - \varphi_0(t - \varepsilon\tau)] \, d\tau \right\|_B^q \, dt \right]^{1/q},$$

Table of spaces defined

Space	Norm	Remarks
C^n		Elements have n continuous derivatives. p. 139.
C_0^n	$\max\limits_{q \in V} \lvert \varphi(q) \rvert$ (if $n = 0$)	Elements have n continuous derivatives and compact support. p. 139.
$\mathscr{C}^n$		$\varphi \in \mathscr{C}^n$ if $\varphi \in C^n$ and $\nabla \cdot \varphi = 0$. p. 140.
$\mathscr{C}_0^n$		$\varphi \in \mathscr{C}_0^n$ if $\varphi \in C_0^n$ and $\nabla \cdot \varphi = 0$. p. 140.
$\mathscr{C}_*^n$		$\varphi \in \mathscr{C}_*^n$ if $\varphi \in \mathscr{C}^n(V) \cap C^0(\overline{V})$ and $\varphi = 0$ on ∂V. p. 148.
L^p	$\lVert \varphi \rVert_p = (\int_V \lvert \varphi \rvert^p \, dq)^{1/p}$	C_0^∞ is dense. If $p = 2$, scalar product is denoted by (φ, ψ). p. 140.
L^∞	$\lVert \varphi \rVert_\infty = \operatorname{ess\,sup} \lvert \varphi(q) \rvert$	p. 140.
$\mathscr{L}^2$	$\lVert \varphi \rVert_2$	$\mathscr{C}_0^\infty$ is dense. p. 140.
$H_0^{1,p}$	$\lVert \nabla \varphi \rVert_p = (\int_V \lvert \nabla \varphi \rvert^p \, dq)^{1/p}$	C_0^∞ is dense. If $p = 2$, scalar product is denoted by $(\nabla \varphi, \nabla \psi)$. p. 143.
$\mathscr{H}_0^{1,p}$	$\lVert \nabla \varphi \rVert_p$	$\mathscr{C}_0^\infty$ is dense. p. 143.
$\mathscr{H}^2$	$\lVert P \nabla^2 \varphi \rVert_2 = (\int_V \lvert P \nabla^2 \varphi \rvert^2 \, dq)^{1/2}$	$\mathscr{C}_*^\infty$ is dense. p. 148.
$C_0(I; B)$		Continuous functions $\varphi : t \to \varphi(t)$ defined on I with values in B. p. 149.
$C_0^0(I; B)$		Functions in $C^0(I; B)$ with compact support in I. p. 149.
$C^1(I; B)$		Functions in $C^0(I; B)$ with one continuous derivative with respect to t. p. 149.
$C_0^1(I; B)$		$C^1(I; B) \cap C_0^0(I; B)$. p. 149.
$L^q(I; B)$	$(\int_I \lVert \varphi(t) \rVert_B^q \, dt)^{1/q}$	$C_0^0(I; B)$ and $C_0^1(I; B)$ are dense. p. 150.
$L^\infty(I; B)$	$\operatorname{ess\,sup} \lVert \varphi(t) \rVert_B$	p. 150.
$L^q(I; L^p)$	$\lVert \varphi \rVert_{p,q}$	p. 150.
$L^q(I; H_0^{1,p})$	$\lVert \nabla \varphi \rVert_{p,q}$	p. 150.
$\mathscr{K}(a, T)$	$[\lVert \nabla \varphi(0) \rVert_2^2 + \int_0^T e^{2at} (\lVert \dot{\varphi}(t) \rVert_2^2 + \lVert P \nabla^2 \varphi(t) \rVert_2^2) \, dt]^{1/2}$	$C_0^1([0, T); \mathscr{H}^2)$ is dense. p. 150.

c always denotes a positive constant.

p denotes a real number ≥ 1. $p' = p/(p - 1)$.

again by (3.2). Therefore, lemma 3.1 gives

$$\left[\int_{-\infty}^{\infty} \| \varphi_0(t) - (k_\varepsilon * \varphi_0)(t) \|_B^q \, dt \right]^{1/q}$$
$$\leqslant \int_{-\infty}^{\infty} k(\tau) \left[\int_{-\infty}^{\infty} \| \varphi_0(t) - \varphi_0(t - \varepsilon\tau) \|_B^q \, dt \right]^{1/q} d\tau.$$

k and φ_0 were both chosen to have compact support. Therefore, there is a closed, bounded interval I outside of which $k(\tau)$ is zero. Also, there is another interval I_0, outside of which both $\varphi_0(t)$ and $\varphi_0(t - \varepsilon\tau)$ are both zero when $\tau \in I$. Moreover, $\varphi_0(t)$ is uniformly continuous. Therefore, if ε is small enough,

$$\left[\int_{-\infty}^{\infty} \| \varphi_0(t) - (k_\varepsilon * \varphi_0)(t) \|_B^q \, dt \right]^{1/q}$$
$$\leqslant \int_I k(\tau) \left[\int_{I_0} \| \varphi_0(t) - \varphi_0(t - \varepsilon\tau) \|_B^q \, dt \right]^{1/q} d\tau$$
$$< \frac{\delta}{3} \int_I k(\tau) \, d\tau = \frac{\delta}{3}. \tag{3.8}$$

Adding (3.6), (3.7), and (3.8), we get

$$\left[\int_{-\infty}^{\infty} \| \varphi(t) - (k_\varepsilon * \varphi)(t) \|_B^q \, dt \right]^{1/q} < \delta$$

if ε is small enough. Since δ is arbitrary, this proves lemma 3.2.

CHAPTER 11

The weak solution

1 Definition of the weak solution

IN THIS chapter, we make our first attempt to show that the Navier–Stokes equations have a solution in a more or less arbitrary domain V. To do so, we define in this section what is meant by a weak solution of the equations. Roughly, a weak solution is a function $s(t, q)$ which satisfies an integral equation derived from the Navier–Stokes equations and which would be a solution if only it had enough derivatives for all the terms appearing in the Navier–Stokes equations to make sense. The concept is due to Leray, but it was Eberhard Hopf[1] who first proved that weak solutions exist in general domains V. On the other hand, the proof given in this chapter is due to me.[2]

Until now, we have looked on each component of the velocity vector s as a real valued function of t and q. At this time, it is convenient to change our view, to treat s as a function of t alone, with values in one of the function spaces defined in the last chapter. Accordingly, we write $s : t \rightarrow s(t)$ for the velocity of a flow, and we write the Navier–Stokes equations (7.2.1.) and (7.2.2) in the form

$$\nabla \cdot s(t) = 0, \qquad (1.1)$$

$$\dot{s}(t) + (s(t) \cdot \nabla) s(t) - \nu \nabla^2 s(t) = - \frac{1}{\varrho} \nabla \pi(t) + Q(t). \qquad (1.2)$$

Here, we have written π for the pressure. $\dot{s}$ is the time derivative of s defined in § 10.3.

Equations (1.1) and (1.2) are to be valid for $t > 0$ (and for q in the domain V of the fluid). In addition to (1.1) and (1.2), we want s to satisfy

$$s(t) = 0 \quad \text{for} \quad q \in \partial V, \qquad (1.3)$$

as we saw in § 7.2.

A classical solution of the Navier–Stokes equations is, more or less, a function $s : t \rightarrow s(t)$, satisfying (1.1–3), for which all the terms appearing in the equations are continuous functions of their arguments. This continuity

---

[1] Eberhard Hopf, Über die Anfangswertaufgabe für die hydrodynamischen Grundgleichungen. Math. Nachr. 4 (1951) 213–231.

[2] Marvin Shinbrot, Fractional derivatives of solutions of the Navier–Stokes equations. Archive for rational mechanics and analysis 40 (1971) 139–154.

requirement can be stated most succinctly in terms of the spaces defined in chapter ten. Since second derivatives appear on the left of (1.2), it is convenient to require that a classical solution s take values in C^2. But (1.1) says that $s(t)$ must lie, not only in C^2, but in the smaller space $\mathscr{C}^2$. Moreover, (1.3) gives that $s(t)$ is in $\mathscr{C}^2_*$, so that a classical solution s takes values in $\mathscr{C}^2_*$. The continuity of a classical solution as a function of t then gives that it lies in $C^0((0, \infty); \mathscr{C}^2_*)$. Also, since a derivative with respect to t appears in (1.2), a classical solution must lie in the space $C^1((0, \infty); C^0)$. Thus, we may define a *classical solution of the Navier–Stokes equations* as a function

$$s \in C^0[(0, \infty); \mathscr{C}^2_*] \cap C^1[(0, \infty); C^0], \tag{1.4}$$

satisfying

$$\dot{s}(t) + (s(t) \cdot \nabla)\, s(t) - \nu\nabla^2 s\,(t) = -\frac{1}{p}\, \nabla\pi\,(t) + Q(t). \tag{1.2}$$

Equations (1.1) and (1.3) can be ignored in stating this definition since every function s with values in $\mathscr{C}^2_*$ satisfies (1.1) and (1.3).

Generally, one is interested not only in finding solutions of the Navier–Stokes equations, but in finding a particular solution with a given initial value. If s^0 is a given function, say in C^0, this means that we want to find a solution of (1.2) satisfying

$$s(0) = s^0. \tag{1.5}$$

In order for (1.5) to make sense classically, we require of a *classical solution of the initial value problem* that it satisfy (1.5) and

$$s \in C^0([0, \infty); C^0), \tag{1.6}$$

as well as (1.4) and (1.2).

Although it is not always necessary, in most of what follows we assume the external forces are derivable from a potential, so that

$$Q(t) = \nabla\Omega\,(t).$$

This assumption simplifies (1.2) a little, for the external force term $\nabla\Omega\,(t)$ can be absorbed in the pressure term, and we can write, instead of (1.2)

$$\dot{s}(t) + (s(t) \cdot \nabla)\, s(t) - \nu\nabla^2 s\,(t) = -\frac{1}{p}\, \nabla\pi\,(t). \tag{1.7}$$

Let s be a classical solution of the initial value problem. Let φ be any element of $C^1_0([0, \infty); \mathscr{C}^\infty_0)$. Multiply (1.7) by $\varphi(t)$ and integrate over V. The term

$$\int_V \nabla\pi\,(t) \cdot \varphi(t)\, dq$$

vanishes, because of lemma 10.2.1. Also, integrating by parts, we find that

$$\int_V (s(t) \cdot \nabla) s(t) \cdot \varphi(t) \, dq = - \int_V (s(t) \cdot \nabla) \varphi(t) \cdot s(t) \, dq,$$

while

$$\int_V \dot{s}(t) \cdot \varphi(t) \, dq = \frac{d}{dt} \int_V s(t) \cdot \varphi(t) \, dq - \int_V s(t) \cdot \dot{\varphi}(t) \, dq.$$

Finally, using Green's theorem twice, we find

$$\int_V \nabla^2 s(t) \cdot \varphi(t) \, dq = \int_V s(t) \cdot \nabla^2 \varphi(t) \, dq.$$

Therefore,

$$\int_V [s(t) \cdot \dot{\varphi}(t) + s(t) \cdot \nabla \varphi(t) \cdot s(t) + s(t) \cdot \nabla^2 \varphi(t)] \, dq = \frac{d}{dt} \int_V s(t) \cdot \varphi(t) \, dq.$$

This equation can be integrated with respect to t. Recalling the notation

$$(\varphi, \psi) = \int_V \varphi \cdot \psi \, dq,$$

introduced in § 10.2, we find

$$\int_0^\infty \{(s(t), \dot{\varphi}(t)) + (s(t) \cdot \nabla \varphi(t), s(t)) + \nu(s(t), \nabla^2 \varphi(t))\} \, dt$$
$$= - (s^0, \varphi(0)). \tag{1.8}$$

Here, the fact that φ has compact support, as well as (1.5) and (1.6), have been used.

What we have proved is that any classical solution of the initial value problem (1.5), (1.7) satisfies (1.8) for all $\varphi \in C_0^1 ([0, \infty); \mathscr{C}_0^\infty)$. On the other hand, it is easy to imagine functions s satisfying (1.8) but which are not sufficiently differentiable for the corresponding equation (1.7) to make any sense at all. A function $s : t \to s(t)$ is called a *weak solution* of the initial value problem (1.5), (1.7) if it is in $L^2 ((0, \infty); \mathscr{H}_0^{1,2})$ and satisfies (1.8) for all $\varphi \in C_0^1 ([0, \infty); \mathscr{C}_0^\infty)$. It is not too hard to show that a weak solution that is smooth enough is actually a classical solution if ∂V is also smooth.

In this chapter, we show that (1.5), (1.7) has at least one weak solution for every initial value $s^0 \in \mathscr{L}^2$.

Exercise

1.1 Show that if s is a weak solution satisfying (1.4) and (1.6) then it is a classical solution of the initial value problem (1.5), (1.7).

2 The Lax–Milgram lemma

To prove the existence of a weak solution, we need a result sometimes called the Lax–Milgram lemma. To prove that, we begin with the following well known result.

LEMMA 2.1   *Let H be a Hilbert space with scalar product* $((\cdot, \cdot))$. *Let A be a bounded linear operator on H. Let N be the nullspace of the adjoint of A, and let R be the range of A. Then, $N = R^{\perp}$.*

Proof   Let $f \in N$. Then, if A^* denotes the adjoint of A, $A^*f = 0$. Therefore, given any $g \in H$,

$$0 = ((A^*f, g)) = ((f, Ag)). \tag{2.1}$$

As g varies over H, Ag varies over R, as that (2.1) says $f \in R^{\perp}$. Thus, $N \subset R^{\perp}$.

For the opposite inclusion, let $f \in R^{\perp}$. This means that

$$((f, Ag)) = 0$$

for all $g \in H$. But $((f, Ag)) = ((A^*f, g))$, so that

$$((A^*f, g)) = 0$$

for all $g \in H$. Setting $g = A^*f$, we conclude that $A^*f = 0$. Therefore, $f \in N$, and $R^{\perp} \subset N$. This completes the proof of lemma 2.1.

We can now prove the Lax–Milgram lemma.

LEMMA 2.2   *Let H be a Hilbert space with scalar product $((\cdot, \cdot))$ and norm $\||\cdot\||$. Let A be a bounded linear operator on H such that*

$$((Af, f)) \geqq c \,\||f\||^2 \tag{2.2}$$

for all $f \in H$ and some positive constant c. Then, $R = H$.

Proof   The Schwarz inequality: $((Af, f)) \leqq \||Af\|| \cdot \||f\||$, along with (2.2), gives

$$\||Af\|| \geqq c \,\||f\||$$

for all $f \in H$. A consequence of this is that R, the range of A, is closed. By definition of A^*, we also have $((Af, f)) = ((f, A^*f))$, and therefore

$$((f, A^*f)) \geqq c\||f\||^2.$$

From this and the Schwarz inequality, we conclude

$$\||A^*f\|| \geqq c\||f\||.$$

Thus N, the nullspace of A^*, is $\{0\}$. Lemma 2.1 then gives that R is dense in H. Since R is both closed and dense, the lemma follows.

3 The quantized Navier–Stokes equations

We shall solve the Navier–Stokes equations by first solving a related, simpler set of equations and then taking a limit. The Navier–Stokes equations themselves are

$$\nabla \cdot s(t) = 0 \tag{3.1}$$

$$\dot{s}(t) + s(t) \cdot \nabla s\,(t) - \nu\nabla^2 s\,(t) = -\frac{1}{\varrho}\,\nabla\pi\,(t). \tag{3.2}$$

The derivative $\dot{s}(t)$ appearing in (3.2) was defined in chapter ten by the formula

$$\dot{s}(t) = \lim_{h \to 0} \frac{1}{h}\,[s\,(t + h) - s(t)]$$

or, what is the same, by the formula

$$\dot{s}(t) = \lim_{h \to 0} \frac{1}{h}\,[s(t) - s\,(t - h)].$$

If, in (3.2), we replace the term $\dot{s}(t)$ by an approximation of the form $(1/h)\,[s(t) - s\,(t - h)]$, we may reasonably expect any solutions of the new equation to be close to a solution of (3.1) and (3.2) when h is small. The same thing should be true if we replace the term $s(t) \cdot \nabla s\,(t)$ by something of the form $s\,(t - h) \cdot \nabla s\,(t)$. We begin by studying the equations obtained from (3.1) and (3.2) after these replacements are made. The equations in question are

$$\nabla \cdot s(t) = 0 \tag{3.3}$$

$$s(t) - s\,(t - h) + hs\,(t - h) \cdot \nabla s\,(t) - h\nu\nabla^2 s\,(t) = -\frac{h}{\varrho}\,\nabla\pi\,(t). \tag{3.4}$$

$s(0)$ is given; it is the initial velocity distribution of the fluid. Setting $t = h$ in (3.3) and (3.4) gives

$$\nabla \cdot s(h) = 0 \tag{3.5}$$

$$s(h) - s(0) + hs\,(0) \cdot \nabla s\,(h) - h\nu\nabla^2 s\,(h) = -\frac{h}{\varrho}\,\nabla\pi\,(h). \tag{3.6}$$

Since $s(0)$ is given, equations (3.5) and (3.6) are equations for $s(h)$. Hopefully, they can be solved. If they can be solved, $s(h)$ can then be found. Then, (3.3) and (3.4) with $t = 2h$ become

$$\nabla \cdot s\,(2h) = 0$$

$$s(2h) - s(h) + hs\,(h) \cdot \nabla s\,(2h) - h\nu\nabla^2 s\,(2h) = -\frac{h}{\varrho}\,\nabla\pi\,(2h).$$

These are now equations for $s(2h)$, since $s(h)$ is known. Again, these equations can, hopefully, be solved for $s(2h)$. The process can clearly be continued and, in this way, we can solve successively for $s(h)$, $s(2h)$, ..., $s(kh)$, ..., given that $s(0)$ is known.

As in § 1, let s^0 be the given value of $s(0)$. Write

$$s^k = s(kh), \quad k = 0, 1, \ldots \tag{3.7}$$

and

$$\pi^k = \pi(kh), \quad k = 0, 1, \ldots$$

Then, setting $t =: kh$ in (3.3) and (3.4), we obtain the following equations for s^k:

$$\nabla \cdot s^k = 0 \tag{3.8}$$

$$s^k - s^{k-1} + hs^{k-1} \cdot \nabla s^k - h\nu\nabla^2 s^k = -\frac{h}{\varrho} \nabla \pi^k, \tag{3.9}$$

$k = 1, 2, \ldots$ We call (3.8) and (3.9) the *quantized Navier–Stokes equations*.

Equations (3.8) and (3.9) have the form

$$\nabla \cdot s = 0 \tag{3.10}$$

$$s + h\sigma \cdot \nabla s - h\nu\nabla^2 s = \sigma - \frac{h}{\varrho} \nabla \pi, \tag{3.11}$$

where we have written σ for s^{k-1} and s for s^k. When σ is given, we call s a *weak solution* of (3.10) and (3.11) if

$$s \in \mathcal{H}_0^{1,2}, \tag{3.12}$$

and

$$(s, \varphi) + h(\sigma \cdot \nabla s, \varphi) + h\nu(\nabla s, \nabla\varphi) = (\sigma, \varphi) \tag{3.13}$$

for all $\varphi \in \mathscr{C}_0^\infty$. (3.13) is obtained formally from (3.11) by multiplication by φ and integrating the third term by parts. As in exercise 1.1, one can show that if a solution s of (3.13) satisfying (3.12) is in C^2, then it satisfies (3.10) and (3.11).

LEMMA 3.1 *Let V be bounded, and let $h > 0$. Then, (3.10) and (3.11) have a weak solution for all $\sigma \in \mathscr{L}^2$.*

Proof We first prove the lemma assuming that $\sigma \in \mathscr{C}_0^\infty$. Let s be any element of $\mathscr{H}_0^{1,2}$, and define a linear functional $F_s(\varphi)$ by the formula

$$F_s(\varphi) = (s, \varphi) + h(\sigma \cdot \nabla s, \varphi) + h\nu(\nabla s, \nabla\varphi). \tag{3.14}$$

If $\sigma \in \mathscr{C}_0^\infty$, $F_s(\varphi)$ is defined for all $\varphi \in \mathscr{H}_0^{1,2}$. In fact,

$$|F_s(\varphi)| \leq \|s\|_2 \|\varphi\|_2 + h\|\sigma\|_\infty \|\nabla s\|_2 \|\varphi\|_2 + h\nu \|\nabla s\|_2 \|\nabla\varphi\|_2$$

$$\leq c \|\nabla s\|_2 \|\nabla\varphi\|_2 \tag{3.15}$$

for some positive constant c, by lemma 10.2.7. (3.15) shows that F_s is actually a bounded linear functional on $\mathcal{H}_0^{1,2}$. Therefore, we conclude from the Riesz representation theorem that there is an $s_1 \in \mathcal{H}_0^{1,2}$ such that

$$F_s(\varphi) = (\nabla s_1, \nabla\varphi). \tag{3.16}$$

What we have proved is that for every $s \in \mathcal{H}_0^{1,2}$, there is an $s_1 \in \mathcal{H}_0^{1,2}$ such that the linear functional F_s, defined by (3.14), has the form (3.16). It is easy to see that the transformation from s to s_1 is linear; we write

$$s_1 = As,$$

so that A is a linear operator on $\mathcal{H}_0^{1,2}$, and

$$F_s(\varphi) = (\nabla As, \nabla\varphi).$$

Setting $\varphi = As$, we obtain

$$\|\nabla As\|_2^2 = F_s(As)$$
$$\leqq c\,\|\nabla s\|_2\,\|\nabla As\|_2,$$

by (3.15). Therefore,

$$\|\nabla As\|_2 \leqq c\,\|\nabla s\|_2,$$

and A is bounded.

When $\varphi = s$, the divergence theorem shows that

$$(\sigma\cdot\nabla s, \varphi) = (\sigma\cdot\nabla s, s) = 0.$$

Therefore,

$$(\nabla As, \nabla s) = F_s(s)$$
$$= \|s\|_2^2 + hv\,\|\nabla s\|_2^2$$
$$\geqq hv\,\|\nabla s\|_2^2.$$

Since the left side of this inequality is the scalar product in $\mathcal{H}_0^{1,2}$ of As and s, lemma 2.2 shows that the range of A is all of $\mathcal{H}_0^{1,2}$.

The functional

$$G_\sigma(\varphi) = (v, \psi)$$

is clearly bounded on $\mathcal{H}_0^{1,2}$. In fact,

$$|G_\sigma(\varphi)| \leqq \|\sigma\|_2\,\|\varphi\|_2 \leqq c\,\|\sigma\|_2\,\|\nabla\varphi\|_2,$$

by lemma 10.2.7. Therefore, again by the Riesz representation theorem, there is a function $\sigma_1 \in \mathcal{H}_0^{1,2}$ such that

$$G_\sigma(\varphi) = (\nabla\sigma_1, \nabla\varphi)$$

for all $\varphi \in \mathscr{H}_0^{1,2}$. The range of A is all of $\mathscr{H}_0^{1,2}$, and $\sigma_1 \in \mathscr{H}_0^{1,2}$. Consequently, there is an s $\in \mathscr{H}_0^{1,2}$ such that

$$As = \sigma_1.$$

We show that this function s is a weak solution of (3.10) and (3.11). We already know that s satisfies (3.12). As for (3.13), let $\varphi \in \mathscr{C}_0^\infty$. Then,

$$(\sigma, \varphi) = (\nabla \sigma_1, \nabla \varphi) = (\nabla As, \nabla \varphi) = F_s(\varphi)$$

$$= (s, \varphi) + h(\sigma \cdot \nabla s, \varphi) + h\nu(\nabla s, \nabla \varphi).$$

Since this is just (3.13), the proof that s is a weak solution of (3.10) and (3.11) is complete. It is important to notice that we have proved, not only that (3.13) is valid when $\varphi \in \mathscr{C}_0^\infty$, but that it is also valid whenever $\varphi \in \mathscr{H}_0^{1,2}$, at least when $\sigma \in \mathscr{C}_0^\infty$.

Now, let $\sigma \in \mathscr{L}^2$. Since $\mathscr{C}_0^\infty$ is dense in $\mathscr{L}^2$, there is a sequence $\{\sigma_j\}$ of elements of $\mathscr{C}_0^\infty$ converging to σ in $\mathscr{L}^2$. Let $\{s_j\}$ be the corresponding sequence of weak solutions of (3.10) and (3.11) whose existence has already been proved. Thus, $\{s_j\} \subset \mathscr{H}_0^{1,2}$, and each s_j satisfies

$$(s_j, \varphi) + h(\sigma_j \cdot \nabla s_j, \varphi) + h\nu(\nabla s_j, \nabla \varphi) = (\sigma_j, \varphi) \tag{3.17}$$

for all $\varphi \in \mathscr{H}_0^{1,2}$. In this formula, set $\varphi = s_j$. The result is

$$\|s_j\|_2^2 + h\nu \|\nabla s_j\|_2^2 = (\sigma_j, s_j). \tag{3.18}$$

Since

$$(\sigma_j, s_j) \leqq \|\sigma_j\|_2 \|s_j\|_2 \leqq \tfrac{1}{2}\|\sigma_j\|_2^2 + \tfrac{1}{2}\|s_j\|_2^2,$$

by lemma 10.2.5, (3.18) gives

$$\|s_j\|_2^2 + 2h\nu \|\nabla s_j\|_2^2 \leqq \|\sigma_j\|_2^2. \tag{3.19}$$

This shows that the sequence $\{s_j\}$ is bounded in $\mathscr{H}_0^{1,2}$. Therefore, there is a subsequence converging weakly in $\mathscr{H}_0^{1,2}$. In order not to proliferate subscripts, we call this subsequence $\{s_j\}$ again, and we call its weak limit s. We show that s is a weak solution of (3.10) and (3.11).

We have (3.17) for all $\varphi \in \mathscr{H}_0^{1,2}$ and certainly then for all $\varphi \in \mathscr{C}_0^\infty$. With $\varphi \in \mathscr{C}_0^\infty$, we send j to infinity in (3.17). Lemma 10.2.7 with $p = p_1 = 2$ shows that $\mathscr{H}_0^{1,2} \subset \mathscr{L}^2$, with continuous injection. Therefore, the weak convergence of s_j in $\mathscr{H}_0^{1,2}$ implies its weak convergence in $\mathscr{L}^2$, and to the same limit. Thus,

$$(s_j, \varphi) \to (s, \varphi) \quad \text{as} \quad j \to \infty.$$

We also have

$$(\nabla s_j, \nabla \varphi) \to (\nabla s, \nabla \varphi) \quad \text{as} \quad j \to \infty$$

by definition of weak convergence in $\mathscr{H}_0^{1,2}$. Consider next

$$|(\sigma_j \cdot \nabla s_j, \varphi) - (\sigma \cdot \nabla s, \varphi)| \leq |((\sigma_j - \sigma) \cdot \nabla s_j, \varphi)| + |(\sigma \cdot \nabla (s_j - s), \varphi)|$$

$$\leq \|\sigma_j - \sigma\|_2 \|\nabla s_j\|_2 \|\varphi\|_\infty + |(\sigma \cdot \nabla (s_j - s), \varphi)|.$$

$$\tag{3.20}$$

$\{\|\nabla s_j\|_2\}$ is bounded. Also, $\{\sigma_j\}$ converges (strongly) to σ in $\mathscr{L}^2$. Therefore, the first term on the right of (3.20) goes to zero as j goes to infinity. As for the second term on the right, any product of components of σ and of φ is square integrable, since $\sigma \in \mathscr{L}^2$ and $\varphi \in \mathscr{C}_0^\infty \subset L^\infty$. Since $\{s_j\}$ converges weakly to s in $\mathscr{H}_0^{1,2}$, this implies that the second term on the right of (3.20) goes to zero also. Therefore,

$$(\sigma_j \cdot \nabla s_j, \varphi) \to (\sigma \cdot \nabla s, \varphi) \quad \text{as} \quad j \to \infty.$$

Putting these results together and letting j go to infinity in (3.17), we obtain (3.13). This shows that s is a weak solution of (3.10) and (3.11) and complete the proof of lemma 3.1.

LEMMA 3.2   *Let V be bounded, and let $h > 0$. Let $\sigma \in \mathscr{L}^2$, and let s be the corresponding weak solution of* (3.10) *and* (3.11). *Then,*

$$\|s\|_2^2 + 2hv \|\nabla s\|_2^2 \leq \|\sigma\|_2^2, \tag{3.21}$$

and

$$\|s\|_2^2 + \|s - \sigma\|_2^2 + 2hv \|\nabla s\|_2^2 \leq \|\sigma\|_2^2. \tag{3.22}$$

Proof   As we saw in the proof of lemma 3.1, we may assume there is a sequence $\{\sigma_j\}$ of elements of $\mathscr{C}_0^\infty$ converging strongly to σ in $\mathscr{L}^2$, and a corresponding sequence $\{s_j\}$ of weak solutions of (3.10) and (3.11) converging weakly to s in both $\mathscr{H}_0^{1,2}$ and $\mathscr{L}^2$. Then, (3.18) and (3.19) are satisfied for all j. If, in (3.19), we allow j to go to infinity, the right side goes to $\|\sigma\|_2^2$, while the left side is never less than $\|s\|_2^2 + 2hv \|\nabla s\|_2^2$, because of lemma 10.2.10. Thus, we obtain (3.21).

To prove (3.22), we use (3.18). It is equivalent to

$$(s_j - \sigma_j, s_j) + hv \|\nabla s_j\|_2^2 = 0.$$

Therefore,

$$\|s_j - \sigma_j\|_2^2 + hv \|\nabla s_j\|_2^2 = - (s_j - \sigma_j, \sigma_j) = \|\sigma_j\|_2^2 - (s_j, \sigma_j)$$

$$= \|\sigma_j\|_2^2 - \|s_j\|_2^2 - hv \|\nabla s_j\|_2^2,$$

by (3.18) again. Consequently,

$$\|s_j\|_2^2 + \|s_j - \sigma_j\|_2^2 + 2hv \|\nabla s_j\|_2^2 = \|\sigma_j\|_2^2.$$

Sending j to infinity in this formula gives (3.22).

With these two results in hand, we can prove the main result of this section.

THEOREM 3.3 *Let V be bounded, and let $h > 0$. If $s^0 \in \mathcal{L}^2$, the quantized Navier–Stokes equatnios (3.8) and (3.9) have a solution s^k for all $k = 1, 2, \ldots$ This solution satisfies*

$$\|s^k\|_2^2 + 2h\nu \|\nabla s^k\|_2^2 \leq \|s^0\|_2^2 \tag{3.23}$$

and

$$\sum_{k=1}^{\infty} \|s^k - s^{k-1}\|_2^2 + 2h\nu \sum_{k=1}^{\infty} \|\nabla s^k\|_2^2 \leq \|s^0\|_2^2. \tag{3.24}$$

Proof That the sequence $\{s^k\}$ of solutions exists is just lemma 3.2, for setting $\sigma = s^{k-1}$ and $s = s^k$ in (3.10) and (3.11) reduces then to (3.8) and (3.9). Thus, it remains to prove (3.23) and (3.24). Setting $\sigma = s^{k-1}$ and $s = s^k$ in (3.21), we find

$$\|s^k\|_2^2 + 2h\nu \|\nabla s^k\|_2^2 \leq \|s^{k-1}\|_2^2.$$

An easy inductive argument now gives (3.23). Finally, set $\sigma = s^{k-1}$ and $s = s^k$ in (3.22). We find

$$\|s^k\|_2^2 + \|s^k - s^{k-1}\|_2^2 + 2h\nu \|\nabla s^k\|_2^2 \leq \|s^{k-1}\|_2^2.$$

Summing over k and cancelling equal terms that appear on both sides of the result, we obtain precisely (3.24).

4   The Navier–Stokes equations in a bounded domain

We can now begin to prove that the Navier–Stokes equations have a weak solution if the domain V of the fluid is bounded. Let h be positive, and let s^0 be any element of $\mathcal{L}^2$. Let $\{s^k\}$ be the sequence of solutions of the quantized Navier–Stokes equations whose existence was just proved. Of course, each s^h depends on h. Define a function $s(t; h)$ as follows:

$$s(t; h) = s^k \text{ for } kh \leq t < (k + 1)h, k = 0, 1, \ldots . \tag{4.1}$$

We show that a weak solution of the Navier–Stokes equations can be constructed from the function $s(t; h)$.

LEMMA 4.1 *Let V be bounded, and let $s^0 \in \mathcal{L}^2$. Then, the function $s(t; h)$ defined by (4.1) satisfies*

$$\|s(t; h)\|_2^2 + 2h\nu \|\nabla s(t; h)\|_2^2 \leq \|s^0\|_2^2 \quad \text{for} \quad t \geq h \tag{4.2}$$

and

$$\frac{1}{h} \int_h^{\infty} \|s(t; h) - s(t - h; h)\|_2^2 \, dt + 2\nu \int_h^{\infty} \|\nabla s(t; h)\|_2^2 \, dt \leq \|s^0\|_2^2. \tag{4.3}$$

It is also true that if $\varphi \in C_0^1 ([0, \infty); \mathscr{C}_0^\infty)$, then

$$\int_0^\infty [(s (t; h), \dot{\varphi}(t)) + (s (t; h) \cdot \nabla\varphi (t), s (t; h))$$

$$+ \nu (s (t; h), \nabla^2\varphi (t))] dt = - (s^0, \varphi(0)) + 0(h^{1/2}) \quad as \quad h \downarrow 0. \tag{4.4}$$

Proof (4.2) is just (3.23), of course, rewritten in terms of $s (t; h)$. To prove (4.3), note that

$$\frac{1}{h} \int_h^\infty \|s (t; h) - s (t - h; h)\|_2^2 \, dt + 2\nu \int_h^\infty \|\nabla s (t; h)\|_2^2 \, dt$$

$$= \frac{1}{h} \sum_{k=1}^\infty \int_{kh}^{(k+1)h} \|s^k - s^{k-1}\|_2^2 \, dt + 2\nu \sum_{k=1}^\infty \int_{kh}^{(k+1)h} \|\nabla s^k\|_2^2 \, dt$$

$$= \sum_{k=1}^\infty \|s^k - s^{k-1}\|_2^2 + 2h\nu \sum_{k=1}^\infty \|\nabla s^k\|_2^2 \leqq \|s^0\|_2^2,$$

by (3.24).

It remains to prove (4.4). Since s^k is a weak solution of the quantized Navier–Stokes equations, we have, if $\varphi \in \mathscr{C}_0^\infty$,

$$(s^k - s^{k-1}, \varphi) + h (s^{k-1} \cdot \nabla s^k, \varphi) + h\nu (\nabla s^k, \nabla\varphi) = 0$$

or, integrating by parts,

$$(s^k - s^{k-1}, \varphi) - h (s^{k-1} \cdot \nabla\varphi, s^k) - h\nu (s^k, \nabla^2\varphi) = 0.$$

Using the definition (4.1) of $s (t; h)$, we see that this is the same as

$$(s (t; h) - s (t - h; h), \varphi) - h (s (t - h; h) \cdot \nabla\varphi, s (t; h)) - h\nu (s (t; h), \nabla^2\varphi)$$

$$= 0, \quad t \geqq h. \tag{4.5}$$

If $\varphi \in C_0^1 ([0, \infty); \mathscr{C}_0^\infty)$, then $\varphi(t) \in \mathscr{C}_0^\infty$ for all $t \geqq 0$. Therefore, we may replace φ by $\varphi(t)$ in (4.5). Integration with respect to t, followed by a little manipulation, then leads to

$$\int_0^\infty [(s(t;h),\dot{\varphi}(t)) + (s(t;h) \cdot \nabla\varphi(t), s(t;h)) + \nu(s(t;h),\nabla^2\varphi(t))] dt + (s^0, \varphi(0))$$

$$= (s^0, \varphi(0)) - \frac{1}{h} \int_0^h (s (t; h), \varphi (t + h)) \, dt$$

$$+ \int_0^h [(s (t; h), \dot{\varphi}(t)) + \nu (s (t; h), \nabla^2\varphi (t)) + (s (t; h) \cdot \nabla\varphi (t), s (t; h))] \, dt$$

$$+ \int_h^\infty \left(s (t; h), \dot{\varphi}(t) + \frac{\varphi(t) - \varphi (t + h)}{h} \right) dt$$

$$+ \int_h^\infty ([s (t; h) - s (t - h; h)] \cdot \nabla\varphi (t), s (t; h)) \, dt. \tag{4.6}$$

The right side of (4.6) contains four terms. We estimate them separately. The first is

$$(s^0, \varphi(0)) - \frac{1}{h} \int_0^h (s(t; h), \varphi(t + h)) \, dt = \frac{1}{h} \int_0^h (s^0, \varphi(0) - \varphi(t + h)) \, dt$$

$$= - \frac{1}{h} \int_0^h \int_0^{t+h} (s^0, \dot{\varphi}(\tau)) \, d\tau \, dt.$$

Therefore,

$$\left| (s^0, \varphi(0)) - \frac{1}{h} \int_0^h (s(t; h), \varphi(t + h)) \, dt \right|$$

$$\leq \frac{1}{h} \|s^0\|_2 \max_{0 \leq \tau \leq 2h} \|\dot{\varphi}(\tau)\|_2 \int_0^h \int_0^{t+h} d\tau \, dt \leq ch, \qquad (4.7)$$

where c is independent of h.

Lemma 10.2.2 shows that the second term on the right of (4.6) is bounded by

$$\int_0^h [\|s(t; h)\|_2 \|\dot{\varphi}(t)\|_2 + \nu \|s(t; h)\|_2 \|\nabla^2 \varphi(t)\|_2 + \|s(t; h)\|_2^2 \|\nabla \varphi(t)\|_\infty] \, dt$$

$$\leq \int_0^h [\|s^0\|_2 \max_{0 \leq t \leq h} \{\|\dot{\varphi}(t)\|_2 + \nu \|\nabla^2 \varphi(t)\|_2\}$$

$$+ \|s^0\|_2^2 \max_{0 \leq t \leq h} \|\nabla \varphi(t)\|_\infty] \, dt = ch: \qquad (4.8)$$

by (4.2).

To estimate the third term, notice that φ has compact support. Therefore, $\varphi(t) \equiv 0$ for $t \geq T$, say, and

$$\left\| \dot{\varphi}(t) + \frac{\varphi(t) - \varphi(t + h)}{h} \right\|_2 \leq ch$$

since $\varphi \in C_0^1 ([0, \infty); \mathscr{C}_0^\infty)$. Therefore,

$$\left| \int_h^\infty \left(s(t; h), \dot{\varphi}(t) + \frac{\varphi(t) - \varphi(t + h)}{h} \right) dt \right|$$

$$\leq \int_h^T \|s(t; h)\|_2 \left\| \dot{\varphi}(t) + \frac{\varphi(t) - \varphi(t + h)}{h} \right\|_2 dt$$

$$\leq ch \int_0^T \|s^0\|_2 \, dt = cTh\|s^0\|_2. \qquad (4.9)$$

The last term on the right of (4.6) is the hardest to estimate. We begin by proving that if σ and s are any two elements of $L^2((h, \infty); \mathcal{H}_0^{1,2})$, and if $\varphi \in C_0^1([0, \infty); \mathcal{C}_0^\infty)$, then

$$\int_h^\infty [(\sigma(t) \cdot \nabla\varphi(t), s(t)) + (\sigma(t) \cdot \nabla s(t), \varphi(t))] \, dt = 0. \tag{4.10}$$

(4.10) is clearly correct if σ and s are in $C_0^0((h, \infty); \mathcal{C}_0^\infty)$, for, in that case

$$(\sigma(t) \cdot \nabla\varphi(t), s(t)) + (\sigma(t) \cdot \nabla s(t), \varphi(t)) = \int_V \sigma(t) \cdot \nabla[\varphi(t) \cdot s(t)] \, dq$$

$$= \int_V \nabla \cdot \{\sigma(t) [\varphi(t) \cdot s(t)]\} \, dq = 0,$$

by the divergence theorem. Since $C_0^0((h, \infty); \mathcal{C}_0^\infty)$ is dense in $L^2((h, \infty); \mathcal{H}_0^2)$, the proof of (4.10) will be complete if we can show that the bilinear functional

$$F(\sigma, s) = \int_h^\infty [(\sigma(t) \cdot \nabla\varphi(t), s(t)) + (\sigma(t) \cdot \nabla s(t), \varphi(t))] \, dt$$

is continuous for σ and s in $L^2((h, \infty); \mathcal{H}_0^{1,2})$. But lemma 10.2.2 shows that

$$|F(\sigma, s)| \leq \int_h^\infty [\|\nabla\varphi(t)\|_\infty \|\sigma(t)\|_2 \|s(t)\|_2 + \|\varphi(t)\|_\infty \|\sigma(t)\|_2 \|\nabla s(t)\|_2] \, dt.$$

The quantities $\|\nabla\varphi(t)\|_\infty$ and $\|\varphi(t)\|_\infty$ are bounded, since $\varphi \in C_0^1([0, \infty); \mathcal{C}_0^\infty)$. Therefore, lemma 10.2.7 gives

$$|F(\sigma, s)| \leq c \int_h^\infty \|\nabla\sigma(t)\|_2 \|\nabla s(t)\|_2 \, dt$$

$$\leq c \left(\int_h^\infty \|\nabla\sigma(t)\|_2^2 \, dt\right)^{1/2} \left(\int_h^\infty \|\nabla s(t)\|_2^2 \, dt\right)^{1/2}.$$

The right side of this inequality is just c times the product of the norms of σ and s in $L^2((h, \infty); \mathcal{H}_0^{1,2})$. Thus, the linear functional F is bounded and, therefore, continuous. This completes the proof of (4.10).

Write $\sigma(t) = s(t; h) - s(t - h; h)$ and $s(t) = s(t; h)$ in (4.10). The functions σ and s are in $L^2((h, \infty); \mathcal{H}_0^{1,2})$ because of (4.2). (4.10) shows that the last term on the right of (4.6) is just

$$-\int_h^\infty ([s(t; h) - s(t - h; h)] \cdot \nabla s(t; h), \varphi(t)) \, dt$$

$$= -\sum_{k=1}^\infty \int_{kh}^{(k+1)h} ((s^k - s^{k-1}) \cdot \nabla s^k, \varphi(t)) \, dt.$$

Thus, this term is bounded by the maximum of $\|\varphi(t)\|_\infty$ times

$$\sum_{k=1}^{\infty} \int_{kh}^{(k+1)h} \|s^k - s^{k-1}\|_2 \, \|\nabla s^k\|_2 \, dt = h \sum_{k=1}^{\infty} \|s^k - s^{k-1}\|_2 \, \|\nabla s^k\|_2$$

$$\leq h \left(\sum_{k=1}^{\infty} \|s^k - s^{k-1}\|_2^2 \right)^{1/2} \left(\sum_{k=1}^{\infty} \|\nabla s^k\|_2^2 \right)^{1/2}$$

$$\leq c \, \|s^0\|_2 \, h^{1/2}, \tag{4.11}$$

by (3.24). Using the estimates (4.7), (4.8), (4.9), and (4.11) in (4.6), we obtain (4.4). This completes the proof of lemma 4.1.

LEMMA 4.2   *Let V be bounded, and let $s^0 \in \mathscr{L}^2$. In addition to having the properties described in lemma 4.1, the function $s(t; h)$ satisfies*

$$|(s(t_2; h) - s(t_1; h), \varphi)| \leq c \, (|t_2 - t_1| + h) \tag{4.12}$$

for all $t_1, t_2 \geq 0$ and all $\varphi \in \mathscr{C}_0^\infty$. Here, c is a constant depending only on s^0 and φ.

Proof   It suffices to prove the lemma assuming $t_2 \geq t_1 \geq 0$. Let k and j be integers such that

$$kh \leq t_2 < (k+1)h, \, jh \leq t_1 < (j+1)h. \tag{4.13}$$

If $k = j$, (4.12) is trivial, since (4.1) shows that the left side of (4.12) is zero. Thus, we may assume $k \geq j + 1$. Since $\{s^n\}$ is a sequence of solutions of the quantized Navier-Stokes equations, we have, as in the proof of lemma 4.1,

$$(s^n - s^{n-1}, \varphi) = h \, (s^{n-1} \cdot \nabla \varphi, s^n) + h\nu \, (s^n, \nabla^2 \varphi)$$

whenever $\varphi \in \mathscr{C}_0^\infty$. Summing over n, we find

$$(s(t_2; h) - s(t_1; h), \varphi) = (s^k - s^j, \varphi)$$

$$= \sum_{n=j+1}^{k} (s^n - s^{n-1}, \varphi)$$

$$= h \sum_{n=j+1}^{k} [(s^{n-1} \cdot \nabla \varphi, s^n) + \nu \, (s^n, \nabla^2 \varphi)].$$

Thus,

$$|(s(t_2; h) - s(t_1; h), \varphi)| \leq h(\|\nabla^2 \varphi\|_2 + \|\nabla \varphi\|_\infty) \sum_{n=j+1}^{k} (\|s^{n-1}\|_2 \, \|s^n\|_2 + \nu \|s^n\|_2)$$

$$\leq h \, (\|\nabla^2 \varphi\|_2 + \|\nabla \varphi\|_\infty) \, (\|s^0\|_2^2 + \nu \, \|s^0\|_2) \sum_{n=j+1}^{k} 1$$

$$= ch \, (k - j). \tag{4.14}$$

In the second step here, (3.23) has been used. Because of (4.13), (4.14) implies

$$|(s(t_2; h) - s(t_1; h), \varphi)| \leqq c(t_2 - t_1 + h),$$

and this proves the lemma.

We can now prove the existence of a weak solution of the Navier-Stokes equations.

THEOREM 4.3 *Let V be bounded, and let $s^0 \in \mathscr{L}^2$. Then the Navier-Stokes equations have a weak solution with s^0 as initial value.*

Proof We show first that there is a sequence $\{h_n\}$, decreasing to zero, such that the corresponding sequence $\{s(t; h_n)\}$ converges weakly in $\mathscr{L}^2$ for all $t \geqq 0$. Because of (4.1), we have

$$\|s(t; h)\|_2 \leqq \|s^0\|_2 \quad \text{for all} \quad t \geqq 0. \tag{4.15}$$

Therefore, the family of functions $\{s(t; h)\}$ is bounded in $\mathscr{L}^2$ for all $t \geqq 0$. It follows that for any fixed $t \geqq 0$, it is possible to find a sequence $\{h_n\}$, decreasing to zero, such that $\{s(t, h_n)\}$ converges weakly in $\mathscr{L}^2$. Let $\{t_m\}$ be the sequence of non-negative rationals. For each m, then, there is a sequence $\{h_{mn}\}$ such that $\{h_{mn}\}$ is a subsequence of $\{h_{m-1, n}\}$, while $\{s(t_m; h_{mn})\}$ converges weakly in $\mathscr{L}^2$. Then, the usual diagonal argument shows that $\{s(t, h_{nn})\}$ converges weakly whenever t is rational. To keep the notation simple, we write $h_n = h_{nn}$. We show that this sequence $\{h_n\}$ has the property that the corresponding sequence $\{s(t; h_n)\}$ convereges weakly for all $t \geqq 0$.

To do this, we must show that

$$\lim_{n \to \infty} (s(t; h_n), \varphi) \tag{4.16}$$

exists for all $\varphi \in \mathscr{L}^2$. For this, it suffices to show that the limit (4.16) exists for all $\varphi \in \mathscr{C}_0^\infty$, since the sequence $\{\|s(t; h_n)\|_2\}$ is bounded. Let $\varphi \in \mathscr{C}_0^\infty$, and take $t \geqq 0$. Let t_1 be a non-negative rational. Then, we have

$$|(s(t; h_n)\varphi) - (s(t_1; h_m), \varphi)| \leqq |(s(t_1; h_n), \varphi) - (s(t_1; h_m), \varphi)|$$

$$| \; |(s(t; h_n) - s(t_1; h_n), \varphi)|$$

$$+ |(s(t; h_m) - s(t_1; h_m), \varphi)|$$

$$\leqq |(s(t_1; h_n), \varphi) - (s(t_1; h_m), \varphi)|$$

$$+ c(|t - t_1| + h_n + h_m),$$

by lemma 4.2. The term $c|t - t_1|$ can be made as small as desired by choosing t_1 close enough to t. The term $c(h_n + h_m)$ can be made small by choosing m and n large enough, since $\{h_n\}$ converges to zero. The remaining term on the right is small when m and n are large since $\{h_n\}$ was chosen in such a way that $\{s(t_1; h_n\}$ converges weakly in $\mathscr{L}^2$ when t_1 is rational. All this shows that $\{(s(t; h_n), \varphi)\}$ is Cauchy, and therefore that the limit (4.16) exists when $\varphi \in \mathscr{C}_0^\infty$. Thus, $\{s(t; h_n)\}$ converges weakly in $\mathscr{L}^2$ for all $t \geq 0$.

Let $s(t)$ be the the weak limit of $\{s(t; h_n)\}$ in $\mathscr{L}^2$. Before going on, we remark that

$$\|s(t)\|_2 \leqq \|s^0\|_2, \, t \geqq 0. \tag{4.17}$$

This follows from (4.15) and lemma 10.2.10.

By (4.15),

$$\int_0^T \|s(t; h_n)\|_2^2 \, dt \leqq T \|s^0\|_2^2$$

for any $T > 0$. This shows that the sequence $\{s(t; h_n)\}$ is bounded in $L^2((0, T); \mathscr{L}^2)$. Therefore, the weak compactness of the ball in Hilbert space shows that there is a subsequence of $\{s(t; h_n)\}$ that converges weakly in $L^2((0, T); \mathscr{L}^2)$. Calling this subsequence $\{s(t; h_n)\}$ again, we see that $\{s(t; h_n)\}$ converges weakly in $L^2((0, T); \mathscr{L}^2)$ as well as in $\mathscr{L}^2$ for every $t \geqq 0$.

Next, (4.3) shows that

$$\int_{h_n}^\infty \|\nabla s(t; h_n)\|_2^2 \, dt \leqq \frac{1}{2\nu} \|s^0\|_2^2.$$

Take $\delta > 0$. Since $\{h_n\}$ goes to zero as n goes to infinity, $h_n \leqq \delta$ if n is large enough. Therefore, for n large enough,

$$\int_\delta^T \|\nabla s(t; h_n)\|_2^2 \, dt \leqq \int_{h_n}^\infty \|\nabla s(t; h_n)\|_2^2 \, dt$$

$$\leqq \frac{1}{2\nu} \|s^0\|_2^2. \tag{4.18}$$

This shows that $\{s(t; h_n)\}$ is bounded in $L^2((\delta, T); \mathscr{H}_0^{1,2})$ for any δ and T with $0 < \delta < T < \infty$. Therefore, as before, $\{s(t; h_n)\}$ has a subsequence converging weakly in $L^2((\delta, T); \mathscr{H}_0^{1,1})$. If, again, we call this new subsequence $\{s(t; h_n)\}$, we see that $\{s(t; h_n)\}$ converges weakly in $L^2((\delta, T); \mathscr{H}_0^{1,2})$, in $L^2((0, T); \mathscr{L}^2)$, and in $\mathscr{L}^2$ for every fixed $t \geqq 0$. The weak limit of $\{s(t; h_n)\}$ in all these spaces is, of course, the same function $s(t)$ defined before.

It follows from (4.18) and the weak convergence in $L^2((\delta, T); \mathscr{H}_0^{1,2})$ that

$$\int_\delta^T \|\nabla s(t)\|_2^2 \, dt \leqq \frac{1}{2\nu} \|s^0\|_2^2. \tag{4.19}$$

Our next task is to show that $\{s(t; h_n)\}$ converges *strongly* to $s(t)$ in $L^2((0, T); \mathcal{L}^2)$. Since $\{s(t; h_n)\}$ converges weakly to $s(t)$ in $L^2((\delta, T); \mathcal{H}_0^{1,2})$, the function $s(t; h_n) - s(t)$ lies in $\mathcal{H}_0^{1,2}$ for almost all t in (δ, T). Therefore, lemma 10.2.9 with $\varphi = s(t; h_n) - s(t)$ gives

$$\|s(t; h_n) - s(t)\|_2^2 \leq \sum_{j=1}^{N} |(s(t; h_n) - s(t), \omega_j)|^2 + \varepsilon \|\nabla s(t; h_n) - \nabla s(t)\|_2^2$$

for almost all $t \in (\delta, T)$, and for every $\varepsilon > 0$. Therefore, we have

$$\int_0^T \|s(t; h_n) - s(t)\|_2^2 \, dt \leq \int_0^\delta \|s(t; h_n) - s(t)\|_2^2 \, dt$$

$$+ \sum_{j=1}^{N} \int_\delta^T |(s(t; h_n) - s(t), \omega_j)|^2 \, dt$$

$$+ \varepsilon \int_\delta^T \|\nabla s(t; h_n) - \nabla s(t)\|_2^2 \, dt.$$

Now,

$$\int_0^\delta \|s(t; h_n) - s(t)\|_2^2 \, dt \leq \int_0^\delta [\|s(t; h_n)\|_2 + \|s(t)\|_2)^2 \, dt$$

$$\leq 4\delta\|s^0\|_2^2,$$

by (4.15) and (4.17). Similarly, (4.18) and (4.19) imply

$$\int_\delta^T \|\nabla s(t; h_n) - \nabla s(t)\|_2^2 \, dt \leq \frac{1}{\nu} \|s^0\|_2^2.$$

Consequently,

$$\int_0^T \|s(t; h_n) - s(t)\|_2^2 \, dt$$

$$\leq 4\delta\|s^0\|_2^2 + \frac{\varepsilon}{\nu} \|s^0\|_2^2 + \sum_{j=1}^{N} \int_\delta^T |(s(t; h_n) - s(t), \omega_j)|^2 \, dt.$$

The first two terms on the right here can be made as small as desired by choosing δ and ε small enough. Once δ and ε are fixed, N is also fixed, and the last term on the right can be made as small as desired by taking n large enough, since $\{s(t; h_n)\}$ converges weakly to $s(t)$ in $L^2((\delta, T); \mathcal{L}^2)$. This shows that $\{s(t; h_n)\}$ converges strongly to $s(t)$ in $L^2((0, T); \mathcal{L}^2)$.

We now prove that $s(t)$ is a weak solution of the Navier–Stokes equations. Let $\varphi \in C^1([0, \infty); \mathscr{C}_0^\infty)$. Choose T so large that $\varphi(t) = 0$ for $t > T$. Then, the integral in (4.4) is just an integral over the interval $0 \leq t \leq T$. In (4.4), we replace h by h_n and let n go to infinity. The sum of the first and third terms on the left of (4.4) converges to

$$\int_0^T [(s(t), \dot{\varphi}(t)) + v\,(s(t), \nabla^2\varphi\,(t))]\,dt$$

$$= \int_0^\infty [(s(t), \dot{\varphi}(t)) + v\,(s(t), \nabla^2\varphi\,(t))]\,dt \tag{4.20}$$

since $\{s\,(t;\,h_n)\}$ converges weakly to $s(t)$ in $L^2\,((0,\,T);\,\mathscr{L}^2)$. We now show that the second term on the left of (4.4) converges to

$$\int_0^\infty (s(t) \cdot \nabla\varphi\,(t),\,s(t))\,dt \tag{4.21}$$

as n goes to infinity. For this purpose, consider

$$\left| \int_0^\infty (s\,(t;\,h_n) \cdot \nabla\varphi\,(t),\,s\,(t;\,h_n))\,dt - \int_0^\infty (s(t) \cdot \nabla\varphi\,(t),\,s(t))\,dt \right|$$

$$\leq \int_0^T |([s\,(t;\,h_n) - s(t)] \cdot \nabla\varphi\,(t),\,s\,(t;\,h_n))|\,dt$$

$$+ \int_0^T |(s(t) \cdot \nabla\varphi\,(t),\,s\,(t;\,h_n) - s(t))|\,dt$$

$$\leq \int_0^T \|\nabla\varphi\,(t)\|_\infty\,\|s\,(t;\,h_n) - s(t)\|_2\,[\|s\,(t;\,h_n)\|_2 + \|s(t)\|_2]\,dt$$

$$\leq 2\,\|s^0\|_2 \max_t \|\nabla\varphi\,(t)\|_\infty \int_0^T \|s\,(t;\,h_n) - s(t)\|_2\,dt$$

$$\leq cT^{1/2} \left(\int_0^T \|s\,(t;\,h_n) - s(t)\|_2^2\,dt \right)^{1/2}.$$

The integral here goes to zero as n goes to infinity since, as we have seen, $\{s\,(t;\,h_n)\}$ converges strongly to $s(t)$ in $L^2\,((0,\,T);\,\mathscr{L}^2)$. Thus, the left side of (4.4) converges to the sum of (4.20) and (4.21) as n goes to infinity. The right side of (4.4) clearly goes to $-(s^0,\,\varphi(0))$ as n goes to infinity. Thus, we have

$$\int_0^\infty [(s(t), \dot{\varphi}(t)) + (s(t) \cdot \nabla\varphi\,(t),\,s(t)) + v\,(s(t), \nabla^2\varphi\,(t))]\,dt = -\,(s^0,\,\varphi\,(0)).$$

This is just (1.8). Also, letting δ tend to zero and T to infinity in (4.19) we find

$$\int_0^\infty \|\nabla s\,(t)\|_2^2\,dt \leq \frac{1}{2v}\,\|s^0\|_2^2. \tag{4.22}$$

Therefore, $s \in L^2\,((0, \infty); \mathscr{H}_0^{1,2})$. This, plus the fact that s satisfies (1.8), shows that it is a weak solution and completes the proof of theorem 4.3.

Exercises

3.1 Show that the Navier–Stokes equations have a weak solution for all $s^0 \in \mathscr{L}^2$, whether or not the domain V is bounded. (*Hint.* If V is unbounded, let $\{V_n\}$ be a sequence of bounded domains increasing to V. Let $\{s_n\}$ be a corresponding sequence of weak solutions in the domains $\{V_n\}$. Show that $\{s_n\}$ has a subsequence converging weakly in $L^2\,((0, \infty);$ $\mathscr{H}_0^{1,2})$, and that its weak limit is a weak solution in V.)

3.2 If an external force term Q is present in the Navier–Stokes equations, we say that s is a *weak solution* of the equations with initial value s^0 if

$$s \in L^2\,((0, \infty); \mathscr{H}_0^{1,2})$$

and

$$\int_0^\infty [(s(t), \dot\varphi(t)) + (s(t) . \nabla\varphi\,(t), s(t)) + \nu\,(s(t), \nabla^2\varphi\,(t))]\,dt$$

$$= -\int_0^\infty (Q(t), \varphi(t))\,dt - (s^0, \varphi(0))$$

for all $\varphi \in C_0^1\,([0, \infty); \mathscr{C}_0^\infty)$.

Show that the Navier–Stokes equations have a weak solution whenever $s^0 \in \mathscr{L}^2$ and $Q \in L^2\,((0, T); L^2)$ for all $T > 0$.

3.3 Prove the following lemma, which is due to Leray. Let V be a domain with a sufficiently smooth boundary. Let $\alpha(q)$ be a sufficiently smooth function, defined for $q \in \partial V$, taking values in R^3. Show that there exists a smooth function $a(q)$, with values in R^3, defined on $\overline{V}$, and such that

$$\nabla \cdot a = 0 \quad \text{in} \quad V,$$

$$a(q) = \alpha(q) \quad \text{for} \quad q \in \partial V,$$

$$|(s \cdot \nabla a, s)| \leqq \varepsilon\,\|\nabla s\|_2^2 \text{ for every } s \in \mathscr{H}_0^{1,2} \text{ and any fixed } \varepsilon > 0,$$

and a is identically zero outside a neighborhood of ∂V.

3.4 Let V be an exterior domain with a smooth boundary. Consider the Navier–Stokes equations in the form

$$\dot s + (s . \nabla)\,s - v\nabla^2 s = -\frac{1}{\varrho}\,\nabla\pi, \tag{4.23}$$

$$\nabla \cdot s = 0, \tag{4.24}$$

$$s = 0 \quad \text{on} \quad \partial V, \tag{4.25}$$

$$s \to s^\infty \quad \text{as} \quad |q| \to \infty, \tag{4.26}$$

$$s(0) = s^0. \tag{4.27}$$

Here, s^∞ is a constant vector. Let $\alpha(q) = -s^\infty$ on ∂V, and let $a(q)$ be the function defined in the preceding exercise. Write $s = s^\infty + a + \sigma$. We say that s is a *weak solution* of (4.23–4.27) if

$$\sigma \in L^2\,((0, \infty); \mathscr{H}_0^{1,2})$$

and σ satisfies

$$(s^0 - s^\infty - a, \varphi(0)) + \int_0^\infty [(\sigma(t), \dot{\varphi}(t)) + (\{s^\infty + a + \sigma(t)\} \cdot \nabla\varphi(t), a + \sigma(t))$$

$$- \nu(a + \sigma(t), \nabla^2\varphi(t))] \, dt = 0$$

for all $\varphi \in C_0^1 [0, \infty).\mathscr{C}_0^\infty)$. Justify this definition and prove that a weak solution exists for every constant s^∞ and every $s^0 \in \mathscr{L}^2$.

5 Some properties of weak solutions

The weak solution whose existence we have proved has certain properties that will be needed in our later work. We derive these properties here.

The first of these is called the *energy inequality*, for reasons that § 7.4 should make clear. We state the energy inequality as a theorem.

THEOREM 5.1 *Let V be bounded, and let $s^0 \in \mathscr{L}^2$. Then, the weak solution s whose existence was proved in theorem 4.3 satisfies the energy inequality*

$$\tfrac{1}{2} \|s(t)\|_2^2 + \nu \int_0^t \|\nabla s(\tau)\|_2^2 \, d\tau \leq \tfrac{1}{2}\|s^0\|_2^2. \tag{5.1}$$

Consequently,

$$s \in L^\infty ((0, \infty); \mathscr{L}^2) \cap L^2 ((0, \infty); \mathscr{H}_0^{1,2}). \tag{5.2}$$

Proof In the course of proving theorem 3.3, we showed that

$$\|s^k\|_2^2 + \|s^k - s^{k-1}\|_2^2 + 2h\nu \|\nabla s^k\|_2^2 \leq \|s^{k-1}\|_2^2, \tag{5.3}$$

where $\{s^k\}$ is the sequence of weak solutions of the quantized Navier–Stokes equations. (5.3) clearly implies

$$\|s^k\|_2^2 + 2h\nu \|\nabla s^k\|_2^2 \leq \|s^{k-1}\|_2^2.$$

Summing on k, we find

$$\|s^j\|_2^2 + 2h\nu \sum_{k=1}^j \|\nabla s^k\|_2^2 \leq \|s^0\|_2^2.$$

As in § 4, define a function $s(t; h)$ by the formula

$$s(t; h) = s^j \quad \text{for} \quad jh \leq t < (j + 1)h, \quad j = 0, 1, \ldots$$

Take $\delta > 0$. Then, if $jh \leq t < (j + 1)h$, and if h is small enough, we have

$$\|s(t; h)\|_2^2 + 2\nu \int_\delta^t \|\nabla s(\tau; h)\|_2^2 \, d\tau \leq \|s^j\|_2^2 + 2\nu \sum_{k=1}^j \int_{kh}^{(k+1)h} \|\nabla s^k\|_2^2 \, d\tau$$

$$= \|s^j\|_2^2 + 2h\nu \sum_{k=1}^j \|\nabla s^k\|_2^2 \leq \|s^0\|_2^2. \tag{5.4}$$

We saw in the proof of theorem 4.3 that there is a sequence $\{h_n\}$ such that $\{s(t; h_n)\}$ converges weakly to s, the weak solution of the Navier–Stokes equations, in $L^2((0, \infty); \mathscr{H}_0^{1,2})$ and in $\mathscr{L}^2$ for every fixed $t \geq 0$. Writing h_n for h in (5.4) letting n go to infinity, and then letting δ go to zero, we find (5.1) because of lemma 10.2.10. This completes the proof of theorems 5.1, since (5.2) follows immediately from (5.1).

COROLLARY 5.2   *The weak solution is in* $L^2((0, \infty); L^6)$.

Proof   Lemma 10.2.6 shows that

$$\int_0^\infty \|s(t)\|_6^2 \, dt \leq c \int_0^\infty \|\nabla s(t)\|_2^2 \, dt \leq c\|s^0\|_2^2,$$

by (5.1). The corollary follows from this.

The weak solution of theorem 4.3 does not have much smoothness with respect to t, but it does have some. How much, is the content of our next result.

THEOREM 5.3   *The weak solution* s *of theorem* 4.3 *is continuous as a function of* t *in the weak topology of* $\mathscr{L}^2$. *It is continuous in the strong topology of* $\mathscr{L}^2$ *at* $t = 0$ *and satisfies*

$$\|s(t) - s^0\|_2 \to 0 \quad \text{as} \quad t \downarrow 0. \tag{5.5}$$

Proof   Let $\{s^k\}$ be the sequence of weak solutions of the quantized Navier–Stokes equations, and let

$$s(t; h) = s^k \quad \text{for} \quad kh \leq t < (k + 1) h.$$

Then, as we saw in equation (4.5) and the equation preceding it,

$$\frac{1}{h}(s(t; h) - s(t - h; h), \varphi) = (s(t - h; h) \cdot \nabla\varphi, s(t; h)) + \nu(s(t; h), \nabla^2\varphi),$$

$$t \geq h, \tag{5.6}$$

and

$$\frac{1}{h}(s^k - s^{k-1}, \varphi) = (s^{k-1} \cdot \nabla\varphi, s^k) + \nu(s^k, \nabla^2\varphi), \quad k = 1, 2, \ldots, \tag{5.7}$$

for all $\varphi \in \mathscr{C}_0^\infty$.

Integrating (5.6) we find,

$$\frac{1}{h}\int_h^t (s(\tau; h) - s(\tau - h; h), \varphi) \, d\tau$$

$$= \int_h^t [(s(\tau - h; h) \, \nabla\varphi, s(\tau; h)) + \nu(s(\tau; h), \nabla^2\varphi)] \, d\tau. \tag{5.8}$$

On the other hand, if $kh \leq t < (k + 1)h$, we have

$$(s(t; h), \varphi) - (s^0, \varphi) - \frac{1}{h} \int_h^t (s(\tau; h) - s(\tau - h; h), \varphi) \, d\tau$$

$$= (s(t; h), \varphi) - (s^0, \varphi) - \frac{1}{h} \int_{t-h}^t (s(\tau; h), \varphi) \, d\tau - \frac{1}{h} \int_0^h (s(\tau; h), \varphi) \, d\tau$$

$$= (s^k, \varphi) - (s^0, \varphi) - \frac{1}{h} \int_{t-h}^{kh} (s^{k-1}, \varphi) \, d\tau - \frac{1}{h} \int_{kh}^t (s^k, \varphi) \, d\tau$$

$$- \frac{1}{h} \int_0^h (s^0, \varphi) \, d\tau$$

$$= \frac{(k+1)h - t}{h} (s^k - s^{k-1}, \varphi)$$

$$= [(k+1)h - t] [(s^{k-1} \cdot \nabla\varphi, s^k) + \nu (s^k, \nabla^2\varphi)],$$

by (5.7). Therefore, Hölder's inequality shows that

$$\left| (s(t; h), \varphi) - (s^0, \varphi) - \frac{1}{h} \int_h^t (s(\tau; h) - s(\tau - h; h), \varphi) \, d\tau \right|$$

$$\leq |(k+1)h - t| [\|s^{k-1}\|_2 \|\nabla\varphi\|_\infty \|s^k\|_2 + \nu\|s^k\|_2 \|\nabla^2\varphi\|_2]$$

$$\leq h [\|\nabla\varphi\|_\infty \|s^0\|_2^2 + \nu \|\nabla^2\varphi\|_2 \|s^0\|_2],$$

by (3.23). This result and (5.8) imply

$$(s(t; h), \varphi) = (s^0, \varphi) + \int_h^t [(s(\tau - h; h) \cdot \nabla\varphi, s(\tau; h))$$

$$+ \nu (s(\tau; h), \nabla^2\varphi)] \, d\tau + \mathrm{O}(h) \tag{5.9}$$

as $h \downarrow 0$. We saw in the proof of theorem 4.3 that there is a sequence $\{h_n\}$ decreasing to zero such that $\{s(t; h_n)\}$ converges weakly in $L^2((0, \infty; \mathscr{H}_0^{1,2})$, strongly in $L^2((0, \infty); \mathscr{L}^2)$, and weakly in $\mathscr{L}^2$ for every fixed $t \geq 0$. Setting $h = h_n$ in (5.9) and sending n to infinity, then, we find

$$(s(t), \varphi) = (s^0, \varphi) + \int_0^t [(s(\tau) \cdot \nabla\varphi, s(\tau)) + \nu (s(\tau), \nabla^2\varphi)] \, d\tau \tag{5.10}$$

for all $\varphi \in \mathscr{C}_0^\infty$.

To prove the first part of the theorem, we have to show that $(s(t), \varphi)$ is continuous in t for all $\varphi \in \mathscr{L}^2$. To begin, let $\varphi \in \mathscr{C}_0^\infty$. Then, $(s(t), \varphi)$ is given by (5.10), while the right side of (5.10) is obviously continuous in t. Next, let $\varphi \in \mathcal{L}^2$. Since $\mathscr{C}_0^\infty$ is dense in $\mathcal{L}^2$, we can find a function $\psi \in \mathscr{C}_0^\infty$ such that

$$\|\varphi - \psi\|_2 < \varepsilon$$

for any $\varepsilon > 0$. Let $t_0, t_1 \in [0, \infty]$, and consider

$$|(s(t_0), \varphi) - (s(t_1), \varphi)| \leqq |(s(t_0) - s(t_1), \psi)|$$
$$+ (\|s(t_0)\|_2 + \|s(t_1)\|_2) \|\varphi - \psi\|_2$$
$$\leqq |(s(t_0) - s(t_1), \psi)| + 2\|s^0\|_2 \varepsilon,$$

by (5.1). The second term here can be made as small as desired by choosing ε small enough. Once ε is chosen, ψ can be determined, and with ψ fixed, the first term can be made as small as desired by choosing t_0 and t_1 close enough together since, as we have already seen, $(s(t), \psi)$ is continuous in t if $\psi \in \mathscr{C}_0^\infty$. This shows that $(s(t), \varphi)$ is continuous for all $\varphi \in \mathcal{L}^2$ and that s is continuous in the weak topology of $\mathcal{L}^2$.

To prove the strong continuity at $t = 0$, we only have to prove (5.5). Note first that (5.10) with $t = 0$ gives

$$(s(0), \varphi) = (s^0, \varphi)$$

for all $\varphi \in \mathscr{C}_0^\infty$. Since $\mathscr{C}_0^\infty$ is dense in $\mathcal{L}^2$, this implies

$$s(0) = s^0.$$

Because of lemma 10.2.10, the weak continuity already proved implies that

$$\liminf_{t \downarrow 0} \|s(t)\|_2^2 \geqq \|s(0)\|_2^2 = \|s^0\|_2^2. \tag{5.11}$$

Therefore, it suffices to prove that

$$\limsup_{t \uparrow 0} \|s(t)\|_2^2 = \|s^0\|_2^2, \tag{5.12}$$

for then (5.11) and (5.12) together give

$$\lim_{t \downarrow 0} \|s(t)\|_2^2 = \|s^0\|_2^2,$$

and this, along with the weak continuity already proved, gives (5.5). Suppose, then, that

$$\limsup_{t \downarrow 0} \|s(t)\|_2^2 > \|s^0\|_2^2. \tag{5.13}$$

Therefore, there is a sequence $\{t_n\}$ decreasing to zero such that

$$\|s(t_n)\|_2^2 \geqq \|s^0\|_2^2 + \varepsilon$$

for some $\varepsilon > 0$. Then, the energy inequality (5.1) gives

$$\|s^0\|_2^2 \geq \|s(t_n)\|_2^2 + 2\nu \int_0^{t_n} \|\nabla s\,(\tau)\|_2^2\, d\tau \geq \|s^0\|_2^2 + \varepsilon,$$

and this contradiction shows that (5.13) is impossible.

What makes the point $t = 0$ special is the fact that the energy inequality (5.1) has, on its right hand side, the norm of the solution at $t = 0$. One cannot necessarily shift the time origin to obtain

$$\tfrac{1}{2}\|s(t)\|_2^2 + \nu \int_{t_0}^t \|\nabla s\,(\tau)\|_2^2\, d\tau \leq \tfrac{1}{2}\|s(t_0)\|_2^2, \qquad (5.14)$$

$t \geq t_0$, reasonable though (5.14) may appear. It would be most interesting to know whether (5.14) is generally true, but all we can prove is that it is true for almost all $t_0 \geq 0$. We begin by proving

LEMMA 5.4   *Let V be bounded. Then, there is a sequence $\{h_n\}$, decreasing to zero, such that the corresponding sequence $\{s\,(t;h_n)\}$ satisfies:*

(i)   *for all $t \geq 0$, $\{s\,(t;h_n)\}$ converges weakly to $s(t)$ in $\mathscr{L}^2$;*
(ii) *$\{s\,(t;h_n)\}$ converges strongly to $s(t)$ in $L^2\,((0,\infty);\mathscr{L}^2)$;*
(iii) *$\{s\,(t;h_n)\}$ converges weakly to $s(t)$ in $L^2\,((0,\infty);\mathscr{H}_0^{1,2})$;*
(iv) *for almost every $t_0 \geq 0$, $\{s\,(t_0;h_n)\}$ converges strongly to $s(t_0)$ in $\mathscr{L}^2$.*

Proof   A sequence $\{h_n\}$ such that (i)–(iii) are satisfied was constructed in the course of proving theorem 4.3. But, because of (ii),

$$\int_0^\infty \|s\,(t;h_n) - s(t)\|_2^2\, dt \to 0$$

and, therefore, there is a subsequence $\{h_n'\}$ such that $\|s\,(t;h_n') - s(t)\|_2^2 \to 0$ almost everywhere[1]. The sequence $\{h_n'\}$ satisfies all the conditions of the lemma.

Now, let t_0 be one of the points for which (iv) holds. A proof like that of theorem 5.1 shows that (5.14) holds for all $t \geq t_0$. But (iv) says that the complement of the set from which t_0 was chosen has measure zero. Therefore, we have

THEOREM 5.5   *Let s be a weak solution. Let S_0 be the set of all points with the following property: (5.14) holds for all $t \geq t_0$. Then, the complement of S_0 has measure zero.*

The weak solution is actually a little smoother than theorem 5.3 shows. We

---

[1] See any textbook on real variables, for example, H.L.Royden, *Real analysis*, Macmillan, 1968, § 4.5.

say that a function $s \in L^2 ((0, \infty); L^2)$ has a *fractional derivative of order α* if the function $D_t^\alpha s$ defined by

$$(D_t^\alpha s) (t) = \frac{1}{\Gamma (1 - \alpha)} \frac{d}{dt} \int_0^t \frac{s(\tau)}{(t - \tau)^\alpha} d\tau$$

exists and is in $L^2 ((0, \infty); L^2)$. Here Γ is just the Euler Γ-function. It is easy to see that

$$D_t^0 s = s,$$

and it is not hard to show that if s has a derivative $\dot{s}$ in $L^2 ((0, \infty); L^2)$, then $D_t^1 s$ exists and

$$D_t^1 s = \dot{s}.$$

To say that $D_t^\alpha s$ exists for $0 < \alpha < 1$ is, then, to say more than s is in $L^2 ((0, \infty); L^2)$ and less than s has a derivative in $L^2 ((0, \infty); L^2)$.

In 1959, J.-L. Lions[1] proved that the weak solution has a fractional derivative of order α for every $\alpha < \frac{1}{4}$. Some time later, I improved this result to $\alpha < \frac{1}{3}$, and Lions again improved it to $\alpha < \frac{2}{5}$. Neither of these last results was published. Recently, I showed[2] that the weak solution derived in this chapter has a fractional derivative of order α for every $\alpha < \frac{1}{2}$. For reasons given in my paper, this result seems surely to be best possible unless the weak solution is actually classical.

Exercises

5.1 Prove theorems 4.3 and 5.1 for two dimensional solutions of the Navier–Stokes equations. Then, using exercise 10.2.2, show that in two-dimensions, corollary 5.2 can be strengthened to read

$$s \in L^2 ((0, \infty); L^p) \quad \text{for every} \quad p < \infty.$$

5.2 Generalize (5.10) by showing that if s is a weak solution, then

$$(s(t), \varphi(t)) = (s^0, \varphi (0)) + \int_0^t [(s(\tau), \dot{\varphi}(\tau)) + (s(\tau) \cdot \nabla \varphi (\tau), s(\tau)) + \nu (s(\tau), \nabla^2 \varphi (\tau))] \, d\tau$$

for all $\varphi \in C_0^1 ([0, \infty); \mathscr{C}_0^\infty)$.

---

[1] Sur l'existence de solutions des équations de Navier–Stokes. Comptes Rendus Acad. Sci. Paris 248 (1959) 2847–2849.

[2] *Fractional derivatives. ... Added in proof:* The proof of lemma 5.2 of my paper is wrong, so that the question of whether the weak solution has a fractional derivative of every order less than $\frac{1}{2}$ remains open. My guess is the answer is Yes, but I cannot prove it.

CHAPTER 12

Uniqueness of weak solutions

1 Introduction

IT IS NOT known whether the weak solution constructed in the last chapter is unique. The basic relations that, in the end, provided the existence of the weak solution are the inequalities (4.17) and (4.18), or, what is essentially the same, the energy inequality

$$\tfrac{1}{2} \|s(t)\|_2^2 + \nu \int_0^t \|\nabla s(\tau)\|_2^2 \, d\tau \leq \tfrac{1}{2}\|s^0\|_2^2. \tag{1.1}$$

In theorems 11.5.1 and 11.5.2, this inequality was used to show that

$$s \in L^\infty((0, \infty); \mathscr{L}^2) \cap L^2((0, \infty); L^6).$$

By lemma 10.2.4, therefore, $s \in L^q((0, \infty); L^p)$ whenever

$$2 \leq p \leq 6 \quad \text{and} \quad q = \frac{4p}{3p - 6}. \tag{1.2}$$

It turns out that whenever a weak solution is in a space $L^q(I; L^p)$ for some interval I, the crucial measure of how near it comes to a classical solution is the quantity[1]

$$\frac{3}{p} + \frac{2}{q};$$

the smaller this is, the better the solution is.

When p and q are related by (1.2), we have

$$\frac{3}{p} + \frac{2}{q} = \frac{3}{2},$$

but the critical value is unity. We prove in this chapter that if, for an interval I whose left-hand endpoint is in the set S_0 of theorem 11.5.5, there is a weak solution s in $L^q(I; L^p)$, where

$$\frac{3}{p} + \frac{2}{q} = 1,$$

then all weak solutions on I having the same initial value as s are equal to s everywhere on I. This result is due to Serrin.[2]

---

[1] The coefficient 3 that appears is the dimension of the space; the 2 appears because of the necessity to operate in a Hilbert space at certain points in the argument.

[2] James Serrin, The initial value problem for the Navier–Stokes equations. In Nonlinear Problems, Rudolph E. Langer, ed. University of Wisconsin Press, 1963.

It can be shown that if there is a weak solution satisfying a little more than (1.3), namely,

$$\frac{3}{p} + \frac{2}{q} < 1,$$

then all weak solutions are actually strong and, indeed, are infinitely differentiable with respect to both q and t, but we won't prove this here.[1,2]

2   Uniqueness of classical solutions

The difficulty in the problem lies in the fact that it is *weak* solutions that are being considered. To underscore this fact, we show here that if the solutions are classical, it is quite easy to prove their uniqueness. To this end, let s_1 and s_2 be two classical solutions of the Navier–Stokes equations (11.1.1–3) with the same initial value s^0. Let $s = s_1 - s_2$. Then,

$$s \in C^0\left((0, \infty); \mathscr{C}_*^2\right) \cap C^1\left((0, \infty); C^0\right) \cap C^0\left([0, \infty); C^0\right),$$

and s satisfies

$$\dot{s}(t) + s(t) \cdot \nabla s_1(t) + s_2(t) \cdot \nabla s(t) - \nu \nabla^2 s(t) = -\frac{1}{\varrho} \nabla \pi(t), \quad (2.1)$$

and

$$s(0) = 0. \quad (2.2)$$

If we multiply (2.1) by $s(t)$ and integrate over V, two of the resulting terms vanish. Indeed,

$$\int_V s_2(t) \cdot \nabla s(t) \cdot s(t) \, dq = \frac{1}{2} \int_V s_2(t) \cdot \nabla |s(t)|^2 dq$$

$$= \frac{1}{2} \int_V \nabla \cdot \left[s_2(t) |s(t)|^2\right] dq$$

$$= 0,$$

by the divergence theorem. Similarly,

$$\int_V s(t) \cdot \nabla \pi(t) \, dq = \int_V \nabla \cdot \left[s(t) \pi(t)\right] dq = 0.$$

---

[1] The differentiability with respect to q in the interior of V was first proved by Serrin. On the interior regularity of weak solutions of the Navier–Stokes equations. Archive for rational mechanics and analysis, 9 (1962) 187–195. In the same paper, Serrin proves the differentiability with respect to t, but he requires another condition.

[2] The smoothness with respect to t, as well as smoothness in q up to the boundary of V was proved by Shmuel Kaniel and Marvin Shinbrot, Smoothness of weak solutions of the Navier–Stokes equations. Archive for rational mechanics and analysis, 24 (1967) 302–324.

Therefore, multiplying (2.1) by $s(t)$ and integrating, we obtain, with the aid of Green's theorem,

$$\frac{1}{2}\frac{d}{dt}\|s(t)\|_2^2 + \nu\|\nabla s(t)\|_2^2 = -(s(t)\cdot\nabla s_1(t), s(t)).$$

Consequently,

$$\frac{d}{dt}\|s(t)\|_2^2 \leqq 2\,|\nabla s_1(t)|_\infty\,\|s(t)\|_2^2,$$

which can be written in the form

$$\frac{d}{dt}\left[\|s(t)\|_2^2\,\exp\left\{-2\int_0^t\|\nabla s_1(\tau)\|_\infty\,d\tau\right\}\right] \leqq 0.$$

This means that the function in brackets does not increase. Since (2.2) shows that it vanishes initially, it must always do so. Therefore, $s(t)\equiv 0$, and $s_1(t)\equiv s_2(t)$.

Thus, we have proved that if s_1 and s_2 are classical solutions with

$$\int_0^T\|\nabla s_1(\tau)\|_\infty\,d\tau < \infty, \tag{2.3}$$

then the two solutions are equal on the interval $[0, T]$. It should be noted that the condition (2.3) is a condition on only *one* of the solutions; given (2.3), it follows that $s_2 = s_1$ and as a consequence that s_2 also satisfies (2.3). This asymmetry in the hypotheses required for uniqueness occurs also in Serrin's uniqueness theorem and is most important for our later work.

3   Uniqueness of weak solutions

Let s_1 be a weak solution of the Navier Stokes equations. Then, exercise 11.5.2 shows that s_1 satisfies

$$(s_1(t), \varphi(t)) = (s^0, \varphi(0)) + \int_0^t [(s_1(\tau), \dot\varphi(\tau)) + (s_1(\tau)\cdot\nabla\varphi(\tau), s_1(\tau))$$

$$+ \nu(s_1(\tau), \nabla^2\varphi(\tau))]\,d\tau \tag{3.1}$$

for all $\varphi \in C_0^1([0, \infty); \mathscr{C}_0^\infty)$. Here, of course, $s^0 = s_1(0)$. If $s_1 \in C_0^0([0, \infty); \mathscr{C}_0^\infty)$ then

$$\int_0^t (s_1(\tau), \nabla^2\varphi(\tau))\,d\tau = -\int_0^t (\nabla s_1(\tau), \nabla\varphi(\tau))\,d\tau, \tag{3.2}$$

as Green's theorem shows. But with t fixed, both sides of (3.2) are continuous linear functionals of s_1 for $s_1 \in L^2((0, \infty); \mathscr{H}_0^{1,2})$. Therefore, (3.2) holds for all $s_1 \in L^2 ((0, \infty); \mathcal{H}_0^{1,2})$ and, since any weak solution lies in this space, (3.1) becomes

$$(s_1(t), \varphi(t)) + v \int_0^t (\nabla s_1 (\tau), \nabla \varphi (\tau)) \, d\tau = (s^0, \varphi (0)) + \int_0^t [(s_1(\tau), \dot{\varphi}(\tau))$$
$$+ (s_1(\tau) \cdot \nabla \varphi (\tau), s_1(\tau))] \, d\tau. \quad (3.3)$$

In the following argument, we take t to be a *fixed* positive value of the time. Let k_ε be a mollifier (see § 10.3), and let $\psi \in C_0^1 ([0, \infty); \mathcal{C}_0^\infty)$. Then, the function $\varphi : \tau \to \varphi(\tau)$ defined by

$$\varphi(\tau) = \int_0^t k_\varepsilon (\tau - \sigma) \, \psi(\sigma) \, d\sigma$$

also lies in $C_0^1 ([0, \infty); \mathcal{C}_0^\infty)$ and may be substituted into (3.3). The result is

$$\int_0^t k_\varepsilon (t - \sigma) (s_1(t), \psi(\sigma)) \, d\sigma + v \int_0^t \int_0^t k_\varepsilon (\tau - \sigma) (\nabla s_1 (\tau), \nabla \psi (\sigma)) \, d\sigma \, d\tau$$
$$= \int_0^t k_\varepsilon(-\sigma) (s^0, \psi(\sigma)) \, d\sigma + \int_0^t \int_0^t \dot{k}_\varepsilon (\tau - \sigma) (s_1(\tau), \psi(\sigma)) \, d\sigma \, d\tau$$
$$+ \int_0^t \int_0^t k_\varepsilon (\tau - \sigma) (s_1(\tau) \cdot \nabla \psi (\sigma), s_1(\tau)) \, d\sigma \, d\tau. \quad (3.4)$$

(3.4) holds for all $\psi \in C_0^1 ([0, \infty); \mathcal{C}_0^\infty)$. We now prove that (3.4) actually holds for all $\psi \in L^2 ([0, \infty); \mathcal{H}_0^{1,2})$. To this end, let $F(\psi)$ be the linear functional defined as the difference between the left and the right sides of (3.4). With ε fixed, k_ε is a continuously differentiable, real-valued function with compact support. Therefore, both k_ε and $\dot{k}_\varepsilon$ are bounded for ε fixed, and we find

$$|F(\psi)| \leqq c \left[(\|s_1(t)\|_2 + \|s^0\|_2) \int_0^t \|\psi(\sigma)\|_2 \, d\sigma + \int_0^t \int_0^t \{\|\nabla s_1(\tau)\|_2 \|\nabla \psi (\sigma)\|_2 \right.$$
$$\left. + \|s_1(\tau)\|_2 \|\psi(\sigma)\|_2 + \|s_1(\tau)\|_4^2 \|\nabla \psi (\sigma)\|_2\} \, d\sigma \, d\tau \right],$$

by Hölder's inequality. Lemma 10.2.7 shows that $\|s_1\|_2$ and $\|\psi\|_2$ are bounded by a constant times $\|\nabla s_1\|_2$ and $\|\nabla \psi\|_2$, respectively. Therefore,

$$|F(\psi)| \leqq c \left[\|s_1(t)\|_2 + \|s^0\|_2 + \int_0^t (\|\nabla s_1 (\tau)\|_2 + \|s_1(\tau)\|_4^2) \, d\tau \right] \int_0^t \|\nabla \psi (\sigma)\|_2 \, d\sigma. \quad (3.5)$$

Now s_1, being a weak solution, is in $L^\infty ((0, \infty); \mathscr{L}^2) \cap L^2 ((0, \infty); \mathcal{H}_0^{1,2})$.

Also, lemmas 10.2.4 and 10.2.6 give

$$\|s_1(\tau)\|_4 \leqq \|s_1(\tau)\|_2^{1/4} \|s_1(\tau)\|_6^{3/4}$$

$$\leqq c \|s_1(\tau)\|_2^{1/4} \|\nabla s_1(\tau)\|_2^{3/4}.$$

Therefore, in the notation introduced on p. 150, the quantity in brackets on the right of (3.5) is bounded by

$$\|s_1\|_{2,\infty} + \|s^0\|_2 + \int_0^t (\|\nabla s_1(\tau)\|_2 + c\|s_1\|_{2,\infty}^{1/2} \|\nabla s_1(\tau)\|_2^{3/2})\, d\tau$$

$$\leqq \|s_1\|_{2,\infty} + \|s^0\|_2 + t^{1/2} \|\nabla s_1\|_{2,2} + ct^{1/4} \|s_1\|_{2,\infty}^{1/2} \|\nabla s_1\|_{2,2}^{3/2},$$

by Hölder's inequality. Since t is fixed, the powers of t appearing here are just constants. Also, the energy inequality (11.5.1) shows that both norms of s_1 that appear are bounded by $\|s^0\|_2$. Thus, the bracketed quantity in (3.5) is finite, and we have

$$|F(\psi)| \leqq c \int_0^t \|\nabla\psi(\sigma)\|_2\, d\sigma,$$

where c depends on ε, t, and $\|s^0\|_2$. A last use of Hölder's inequality yields

$$|F(\psi)| \leqq ct^{1/2} \|\nabla\psi\|_{2,2}.$$

This means that $F(\psi)$, the difference between the two sides of (3.4) is continuous for $\psi \in L^2((0,\infty); \mathscr{H}_0^{1,2})$. Because of (3.4), $F(\psi)$ is zero on the subset $C_0^1([0,\infty); \mathscr{C}_0^\infty)$. Therefore, $F(\psi)$ is zero for all $\psi \in L^2((0,\infty); \mathscr{H}_0^{1,2})$, which is only another way of saying that (3.4) holds for all ψ in this last space.

Let s_2 be a second weak solution with the same initial value as s_1. Then, $s_2 \in L^2((0,\infty); \mathscr{H}_0^{1,2})$, and we may set ψ equal to s_2 in (3.4) to obtain

$$\int_0^t k_\varepsilon(t-\sigma)(s_1(t), s_2(\sigma))\, d\sigma + \nu \int_0^t \int_0^t k_\varepsilon(\tau-\sigma)(\nabla s_1(\tau), \nabla s_2(\sigma))\, d\sigma\, d\tau$$

$$= \int_0^t k_\varepsilon(-\sigma)(s^0, s_2(\sigma))\, d\sigma + \int_0^t \int_0^t \dot{k}_\varepsilon(\tau-\sigma)(s_1(\tau), s_2(\sigma))\, d\sigma\, d\tau$$

$$+ \int_0^t \int_0^t k_\varepsilon(\tau-\sigma)(s_1(\tau)\cdot\nabla s_2(\sigma), s_1(\tau))\, d\sigma\, d\tau. \tag{3.6}$$

(3.6) holds for any two weak solutions s_1 and s_2; therefore, it holds when s_1 and s_2 are interchanged. Interchanging s_1 and s_2, the nonlinear term becomes

$$\int_0^t \int_0^t k_\varepsilon(\tau-\sigma)(s_2(\tau)\cdot\nabla s_1(\sigma), s_2(\tau))\, d\sigma\, d\tau. \tag{3.7}$$

We show now that this is equal to

$$-\int_0^t \int_0^t k_\varepsilon(\tau-\sigma)(s_2(\tau)\cdot\nabla s_2(\tau), s_1(\sigma))\, d\sigma\, d\tau. \tag{3.8}$$

For this purpose, define a trilinear functional F by the formula

$$F(\varphi, \psi, \chi) = \int_0^t \int_0^t k_\varepsilon(\tau - \sigma) \left[(\varphi(\tau) \cdot \nabla \psi(\sigma), \chi(\tau)) + (\varphi(\tau) \cdot \nabla \chi(\tau), \psi(\sigma)) \right] d\sigma \, d\tau.$$

For fixed ε, k_ε is bounded. Therefore, Hölder's inequality gives

$$
\begin{aligned}
|F(\varphi, \psi, \chi)| &\leq c \int_0^t \|\varphi(\tau)\|_4 \, \|\chi(\tau)\|_4 \, d\tau \int_0^t \|\nabla \psi(\sigma)\|_2 \, d\sigma \\
&+ c \int_0^t \|\varphi(\tau)\|_3 \, \|\nabla \chi(\tau)\|_2 \, d\tau \int_0^t \|\psi(\sigma)\|_6 \, d\sigma \\
&\leq c \int_0^t \|\nabla \varphi(\tau)\|_2 \, \|\nabla \chi(\tau)\|_2 \, d\tau \int_0^t \|\nabla \psi(\sigma)\|_2 \, d\sigma,
\end{aligned}
$$

by lemma 10.2.6. Hölder's inequality now gives

$$|F(\varphi, \psi, \chi)| \leq c t^{1/2} \|\nabla \varphi\|_{2,2} \, \|\nabla \chi\|_{2,2} \, \|\nabla \psi\|_{2,2},$$

so that F is continuous on

$$L^2\left((0, \infty); \mathscr{H}_0^{1,2}\right) \times L^2\left((0, \infty); \mathscr{H}_0^{1,2}\right) \times L^2\left((0, \infty); \mathscr{H}_0^{1,2}\right).$$

On the other hand, on the dense subset

$$C_0^1\left([0, \infty); \mathscr{C}_0^\infty\right) \times C_0^1\left([0, \infty); \mathscr{C}_0^\infty\right) \times C_0^1\left([0, \infty); \mathscr{C}_0^\infty\right),$$

we have

$$
\begin{aligned}
(\varphi(\tau) \cdot \nabla \psi(\sigma), \chi(\tau)) + (\varphi(\tau) \cdot \nabla \chi(\tau), \psi(\sigma)) &= \int_V \varphi(\tau) \cdot \nabla \left[\psi(\sigma) \cdot \chi(\tau) \right] dq \\
&= \int_V \nabla \cdot \left\{ \varphi(\tau) \left[\psi(\sigma) \cdot \chi(\tau) \right] \right\} dq \\
&= 0,
\end{aligned}
$$

by the divergence theorem. Thus, F in zero on the dense subset, and, therefore, everywhere. Putting $\varphi = \chi = s_2$, $\psi = s_1$ in F, we find that (3.7) and (3.8) are equal. Now, interchange s_1 and s_2 in (3.6) and use the equality of (3.7) and (3.8). The result is

$$
\begin{aligned}
\int_0^t k_\varepsilon(t - \sigma) \, (s_2(t), s_1(\sigma)) \, d\sigma &+ \nu \int_0^t \int_0^t k_\varepsilon(\tau - \sigma) \, (\nabla s_2(\tau), \nabla s_1(\sigma)) \, d\sigma \, d\tau \\
= \int_0^t k_\varepsilon(-\sigma) \, (s^0, s_1(\sigma)) \, d\sigma &+ \int_0^t \int_0^t \dot{k}_\varepsilon(\tau - \sigma) \, (s_2(\tau), s_1(\sigma)) \, d\sigma \, d\tau \\
&- \int_0^t \int_0^t k_\varepsilon(\tau - \sigma) \, (s_2(\tau) \cdot \nabla s_2(\tau), s_1(\sigma)) \, d\sigma \, d\tau. \quad (3.9)
\end{aligned}
$$

Consider the second term on the right of (3.9). Interchanging σ and τ, we find

$$\int_0^t \int_0^t \dot{k}_\varepsilon (\tau - \sigma) (s_2(\tau), s_1(\sigma))\, d\sigma\, d\tau = \int_0^t \int_0^t \dot{k}_\varepsilon (\sigma - \tau) (s_1(\tau), s_2(\sigma))\, d\sigma\, d\tau.$$

$$(3.10)$$

k_ε is any mollifier. The definition of a mollifier does not preclude the possibility of choosing k_ε as an even function. If such a choice is made, $\dot{k}_\varepsilon$ becomes an odd function, so that (3.10) gives

$$\int_0^t \int_0^t \dot{k}_\varepsilon (\tau - \sigma) (s_2(\tau), s_1(\sigma))\, d\sigma\, d\tau = -\int_0^t \int_0^t \dot{k}_\varepsilon (\tau - \sigma) (s_1(\tau), s_2(\sigma))\, d\sigma\, d\tau.$$

$$(3.11)$$

We now add (3.6) and (3.9). (3.11) shows that the terms involving the derivative of k_ε cancel. Also, since k_ε is even, the second terms on the left of (3.6) and (3.9) collect. Therefore, we find

$$\int_0^t k_\varepsilon (t - \sigma) [(s_1(t), s_2(\sigma)) + (s_2(t), s_1(\sigma))]\, d\sigma$$

$$+ 2\nu \int_0^t \int_0^t k_\varepsilon (\tau - \sigma) (\nabla s_1 (\tau), \nabla s_2 (\sigma))\, d\sigma\, d\tau$$

$$= \int_0^t k_\varepsilon(\sigma) (s^0, s_1 (\sigma) + s_2(\sigma))\, d\sigma + \int_0^t \int_0^t k_\varepsilon (\tau - \sigma) [(s_1(\tau) \cdot \nabla s_2 (\sigma), s_1(\tau))$$

$$- (s_2(\tau) \cdot \nabla s_2 (\tau), s_1(\sigma))]\, d\sigma\, d\tau.$$

$$(3.12)$$

With (3.12) in hand, we can prove

LEMMA 3.1   *Let V be bounded. Let s_1 and s_2 be any two weak solutions of the Navier–Stokes equations with $s_1(0) = s_2(0) = s^0$. If $s_1 \in L^q ((0, T); L^p)$, where*

$$\frac{3}{p} + \frac{2}{q} = 1,$$

$$(3.13)$$

then

$$(s_1(t), s_2(t)) = \|s^0\|_2^2 + \int_0^t \{([s_1(\tau) - s_2(\tau)] \cdot \nabla s_2 (\tau), s_1(\tau))$$

$$- 2\nu (\nabla s_1 (\tau), \nabla s_2 (\tau))\}\, d\tau$$

$$(3.14)$$

for all $t \in [0, T)$.

Proof   Let t be a fixed number in $[0, T)$. Define $s_1(\tau)$ and $s_2(\tau)$ as zero for τ outside the interval $[0, t]$. Then, (3.12) can be written in the form

$$\int_{-\infty}^{\infty} k_\varepsilon (t - \sigma) [(s_1(t), s_2(\sigma)) + (s_2(t), s_1(\sigma))] \, d\sigma$$

$$+ 2\nu \int_{-\infty}^{\infty} \int_{-\infty}^{\infty} k_\varepsilon (\tau - \sigma) (\nabla s_1 (\tau), \nabla s_2 (\sigma)) \, d\sigma \, d\tau$$

$$= \int_{-\infty}^{\infty} k_\varepsilon(\sigma)(s^0, s_1(\sigma) + s_2(\sigma)) \, d\sigma + \int_{-\infty}^{\infty} \int_{-\infty}^{\infty} k_\varepsilon(\tau - \sigma) [(s_1(\tau) \cdot \nabla s_2(\sigma), s_1(\tau))$$

$$- (s_2(\tau) \cdot \nabla s_2 (\tau), s_1(\sigma))] \, d\sigma \, d\tau. \tag{3.15}$$

The argument proceeds by letting ε go to zero in (3.15) and using lemma 10.3.2. To show that the lemma applies, we examine the terms in (3.15) one at a time. To begin with the easiest, consider

$$\left| \int_{-\infty}^{\infty} \int_{-\infty}^{\infty} k_\varepsilon (\tau - \sigma) (\nabla s_1 (\tau), \nabla s_2 (\sigma)) \, d\sigma \, d\tau - \int_{-\infty}^{\infty} (\nabla s_1 (\tau), \nabla s_2 (\tau)) \, d\tau \right|$$

$$= \left| \int_{-\infty}^{\infty} (\nabla s_1 (\tau), \nabla [(k_\varepsilon * s_2) (\tau) - s_2(\tau)]) \, d\tau \right|$$

$$\leqq \int_{-\infty}^{\infty} \| \nabla s_1 (\tau) \|_2 \, \| \nabla [(k_\varepsilon * s_2) (\tau) - s_2(\tau)] \|_2 \, d\tau$$

$$\leqq \| \nabla s_1 \|_{2,2} \left[\int_{-\infty}^{\infty} \| \nabla (k_\varepsilon * s_2) (\tau) - \nabla s_2 (\tau) \|_2^2 \, d\tau \right]^{1/2}$$

$$\to 0 \quad \text{as} \quad \varepsilon \downarrow 0,$$

by lemma 10.3.2, since $s_2 \in L^2 ((-\infty, \infty); \mathscr{H}_0^{1,2})$.

Next, consider the first term on the left of (3.15). k_ε is defined by the formula

$$k_\varepsilon(\tau) = \frac{1}{\varepsilon} k \left(\frac{\tau}{\varepsilon} \right).$$

Take $k(\tau)$ to be zero outside the interval $|\tau| < 1$. Then, $k_\varepsilon (\tau) = 0$ for $|\tau| \geqq \varepsilon$, and

$$\int_0^\varepsilon k_\varepsilon(\tau) \, d\tau = \frac{1}{2} \int_{-\infty}^{\infty} k_\varepsilon(\tau) \, d\tau = \frac{1}{2}, \tag{3.16}$$

by definition of a mollifier. Since $s_2(\tau)$ is (by its redefinition) zero outside the interval $[0, t]$, the first term on the left of (3.15) equals

$$\int_0^t k_\varepsilon (t - \sigma) (s_1(t), s_2(\sigma)) \, d\sigma = \int_0^\varepsilon k_\varepsilon(\tau) (s_1(t), s_2 (t - \tau)) \, d\tau,$$

if $t > \varepsilon$. Therefore, in view of (3.16),

$$\left| \int_{-\infty}^{\infty} k_\varepsilon(t - \sigma) (s_1(t), s_2(\sigma)) \, d\sigma - \tfrac{1}{2}(s_1(t), s_2(t)) \right|$$

$$= \left| \int_0^\varepsilon k_\varepsilon(\tau) (s_1(t), s_2(t - \tau) - s_2(t)) \, d\tau \right|$$

$$\leq \tfrac{1}{2} \max_{0 \leq \tau \leq \varepsilon} |(s_1(t), s_2(t - \tau) - s_2(t))|.$$

This maximum goes to zero as ε does, since s_2 is continuous in the weak topology of $\mathscr{L}^2$ (theorem 11.5.3). Thus,

$$\int_{-\infty}^{\infty} k_\varepsilon(t - \sigma) (s_1(t), s_2(\sigma)) \, d\sigma \to \tfrac{1}{2}(s_1(t), s_2(t)) \quad \text{as} \quad \varepsilon \downarrow 0.$$

The same argument applies to the other term on the left of (3.13) and to the first term on the right.

The next term on the right is

$$\int_{-\infty}^{\infty} \int_{-\infty}^{\infty} k_\varepsilon(\tau - \sigma) (s_1(\tau) \cdot \nabla s_2(\sigma), s_1(\tau)) \, d\sigma \, d\tau.$$

It is to evaluate this terms as $\varepsilon \downarrow 0$ that we use the hypothesis $s_1 \in L^q((0, T); L^p)$. Consider

$$\int_{-\infty}^{\infty} \int_{-\infty}^{\infty} k_\varepsilon(\tau - \sigma)(s_1(\tau) \cdot \nabla s_2(\sigma), s_1(\tau)) \, d\sigma \, d\tau - \int_{-\infty}^{\infty} (s_1(\tau) \cdot \nabla s_2(\tau), s_1(\tau)) \, d\tau$$

$$= \int_{-\infty}^{\infty} (s_1(\tau) \cdot \nabla [(k_\varepsilon * s_2)(\tau) - s_2(\tau)], s_1(\tau)) \, d\tau.$$

We have, using lemma 10.2.2,

$$|(s_1(\tau) \cdot \nabla [(k_\varepsilon * s_2)(\tau) - s_2(\tau)], s_1(\tau))|$$

$$\leq \|s_1(\tau)\|_p \, \|s_1(\tau)_{2p/(p-2)} \, \|\nabla [k * s_2)(\tau) - s_2(\tau)]\|_2. \tag{3.17}$$

Therefore, again using lemma 10.2.2,

$$\left| \int_{-\infty}^{\infty} \int_{-\infty}^{\infty} k_\varepsilon(\tau - \sigma)(s_1(\tau) \cdot \nabla s_2(\sigma), s_1(\tau) \, d\sigma \, d\tau - \int_{-\infty}^{\infty} (s_1(\tau) \cdot \nabla s_2(\tau), s_1(\tau)) \, d\tau \right|$$

$$\leq \|s_1\|_{p,q} \, \|s_1\|_{2p/(p-2),\, 2q/(q-2)} \left[\int_{-\infty}^{\infty} \|\nabla (k_\varepsilon * s_2)(\tau) - s_2(\tau)\|_2^2 \, d\tau \right]^{1/2}.$$

$$\tag{3.18}$$

Lemma 10.2.4 shows that

$$\|s_1(\tau)\|_{2p/(p-2)} \leq \|s_1(\tau)\|_2^{1-3/p} \, \|s_1(\tau)\|_6^{3/p} \leq c \, \|s_1(\tau)\|_2^{2/q} \, \|\nabla s_1(\tau)\|_2^{3/p},$$

by (3.13) and lemma 10.2.6. Therefore, another use of (3.13) shows that

$$\|s_1\|_{2p/(p-2),\, 2q/(q-2)} \leqq c \left(\int_{-\infty}^{\infty} \|s_1(\tau)\|_2^{4/(q-2)} \|\nabla s_1(\tau)\|_2^2 \, d\tau \right)^{(q-2)/2q}$$

$$\leqq c \, \|s_1\|_{2,\infty}^{2/q} \|\nabla s_1\|_{2,2}^{3/p}.$$

Use of this inequality in (3.18) results in

$$\left| \int_{-\infty}^{\infty} \int_{-\infty}^{\infty} k_\varepsilon(\tau - \sigma)(s_1(\tau) \cdot \nabla s_2(\sigma), s_1(\tau)) \, d\sigma \, d\tau - \int_{-\infty}^{\infty} (s_1(\tau) \cdot \nabla s_2(\tau), s_1(\tau)) \, d\tau \right|$$

$$\leqq c \| s_1 \|_{p,q} \|s_1\|_{2,\infty}^{2/q} \|\nabla s_1\|_{2,2}^{3/p} \left[\int_{-\infty}^{\infty} \|\nabla (k_\varepsilon * s_2)(\tau) - \nabla s_2(\tau)\|_2^2 \, d\tau \right]^{1/2}.$$

$\|s_1\|_{p,q} < \infty$ by hypothesis, while $\|s_1\|_{2,\infty}$ and $\|\nabla s_1\|_{2,2}$ are finite because of the energy inequality (11.5.1). Lemma 10.3.2 therefore shows that

$$\int_{-\infty}^{\infty} \int_{-\infty}^{\infty} k_\varepsilon(\tau - \sigma)(s_1(\tau) \cdot \nabla s_2(\sigma), s_1(\tau)) \, d\sigma \, d\tau$$

$$\to \int_{-\infty}^{\infty} (s_1(\tau) \cdot \nabla s_2(\tau), s_1(\tau)) \, d\tau \quad \text{as} \quad \varepsilon \downarrow 0.$$

The last term we have to consider is

$$\int_{-\infty}^{\infty} \int_{-\infty}^{\infty} k_\varepsilon(\tau - \sigma)(s_2(\tau) \cdot \nabla s_2(\tau), s_1(\sigma)) \, d\sigma \, d\tau.$$

The argument here is similar to the last one, but instead of (3.17), we use the inequality

$$|(s_2(\tau) \cdot \nabla s_2(\tau), (k_\varepsilon * s_1)(\tau) - s_1(\tau))|$$

$$\leqq \|s_2(\tau) \cdot \nabla s_2(\tau)\|_{p'} \|(k_\varepsilon * s_1)(\tau) - s_1(\tau)\|_p,$$

and we estimate $\|s_2(\tau) \cdot \nabla s_2(\tau)\|_{p'}$ as follows. By lemma 10.2.2,

$$\|s_2(\tau) \cdot \nabla s_2(\tau)\|_{p'} \leqq \|s_2(\tau)\|_{2p/(p-2)} \|\nabla s_2(\tau)\|_2 \leqq c \, \|s_2(\tau)\|_2^{2/q} \|\nabla s_2(\tau)\|_2^{2/q'},$$

by (3.13) and lemmas 10.2.4 and 10.2.6. Therefore,

$$\int_{-\infty}^{\infty} \|s_2(\tau) \cdot \nabla s_2(\tau)\|_{p'}^{q'} \, d\tau \leqq c \, \|s_2\|_{2,\infty}^{2/(q-1)} \|\nabla s_2\|_{2,2}^2.$$

The argument now goes through as before to show that

$$\int_{-\infty}^{\infty} \int_{-\infty}^{\infty} k_\varepsilon(\tau - \sigma)(s_2(\tau) \cdot \nabla s_2(\tau), s_1(\sigma)) \, d\sigma \, d\tau$$

$$\to \int_{-\infty}^{\infty} (s_2(\tau) \cdot \nabla s_2(\tau), s_1(\tau)) \, d\tau \quad \text{as} \quad \varepsilon \downarrow 0.$$

Putting all these results together, we see that letting ε decrease to zero in (3.15) results in

$$(s_1(t), s_2(t))$$
$$= \|s^0\|_2^2 + \int_{-\infty}^{\infty} \{([s_1(\tau) - s_2(\tau)] \cdot \nabla s_2(\tau), s_1(\tau)) - 2\nu \, (\nabla s_1 \, (\tau), \nabla s_2 \, (\tau))\} \, d\tau.$$

Since $s_1(\tau)$ and $s_2(\tau)$ have been defined to be zero outside $[0, t]$, this is the same as (3.14).

LEMMA 3.2   *Take $t > 0$, fixed. Let $s \in L^\infty \, ((0, t); \mathscr{L}^2) \cap L^2 \, ((0, t); \mathscr{H}_0^{1,2})$, and let $\varphi \in L^2 \, ((0, t); \mathscr{H}_0^{1,2}) \cap L^q \, ((0, t); L^p)$, where*

$$\frac{3}{p} + \frac{2}{q} = 1.$$

Then,

$$\int_0^t (s(\tau) \cdot \nabla \varphi \, (\tau), \varphi(\tau)) \, d\tau = 0.$$

Proof   Define a bilinear functional F on the space $L^2 \, ((0, t); \mathscr{H}_0^{1,2}) \times L^q \, ((0, t); L^p)$ by the formula

$$F \, (\varphi, \psi) = \int_0^t (s(\tau) \cdot \nabla \varphi \, (\tau), \psi(\tau)) \, d\tau.$$

F is bounded. Indeed, using the type of argument we have seen several times now, one can show that

$$|F \, (\varphi, \psi)| \leqq c \, \|s\|_{2,\infty}^{2/q} \, \|\nabla s\|_{2,2}^{3/p} \, \|\nabla \varphi\|_{2,2} \, \|\psi\|_{p,q}.$$

On the other hand, on the dense subset $C_0^0 \, ((0, t); \mathscr{C}_0^\infty) \times C_0^0 \, ((0, t); \mathscr{C}_0^\infty)$, the divergence theorem shows that

$$F \, (\varphi, \varphi) = 0.$$

The lemma follows from these two facts.

Now, we can prove

THEOREM 3.3   *Let V be bounded. Let s_1 and s_2 be weak solutions of the Navier–Stokes equations with $s_1(0) = s_2(0)$. If $s_1 \in L^q \, ((0, T); L^p)$, where*

$$\frac{3}{p} + \frac{2}{q} = 1, \tag{3.19}$$

then $s_2(t) = s_1(t)$ for every $t \in [0, T]$.

Proof   Lemma 11.5.1 shows that s_1 and s_2 satisfy

$$\|s_1(t)\|_2^2 + 2\nu \int_0^t \|\nabla s_1 \, (\tau)\|_2^2 \, d\tau \leqq \|s^0\|_2^2 \tag{3.20}$$

and

$$\|s_2(t)\|_2^2 + 2\nu \int_0^t \|\nabla s_2 \, (\tau)\|_2^2 \, d\tau \leqq \|s^0\|_2^2, \tag{3.21}$$

where we have written s^0 for the common value of $s_1(0)$ and $s_2(0)$. Write

$$s = s_2 - s_1.$$

Adding (3.20) and (3.21) and subtracting twice (3.14) from the result, we find

$$\|s(t)\|_2^2 + 2\nu \int_0^t \|\nabla s(\tau)\|_2^2 \, d\tau \leq 2 \int_0^t (s(\tau) \cdot \nabla s_2(\tau), s_1(\tau)) \, d\tau.$$

But lemma 3.2 shows that

$$\int_0^t (s(\tau) \cdot \nabla s_1(\tau), s_1(\tau)) \, d\tau = 0.$$

Therefore,

$$\|s(t)\|_2^2 + 2\nu \int_0^t \|\nabla s(\tau)\|_2^2 \, d\tau \leq 2 \int_0^t (s(\tau) \cdot \nabla s(\tau), s_1(\tau)) \, d\tau. \quad (3.22)$$

We estimate the right side of (3.22) as follows.

$$|(s(\tau) \cdot \nabla s(\tau), s_1(\tau))| \leq \|s(\tau)\|_{2p/(p-2)} \|\nabla s(\tau)\|_2 \|s_1(\tau)\|_p$$
$$\leq c \|s(\tau)\|_2^{2/q} \|\nabla s(\tau)\|_2^{1+3/p} \|s_1(\tau)\|_p,$$

by lemmas 10.2.2, 10.2.4, and 10.2.6. But (3.19) shows that $1 + 3/p = 2/q'$. Therefore,

$$\int_0^t |(s(\tau) \cdot \nabla s(\tau), s_1(\tau))| \, d\tau$$

$$\leq c \int_0^t \|s(\tau)\|_2^{2/q} \|\nabla s(\tau)\|_2^{2/q'} \|s_1(\tau)\|_p \, d\tau$$

$$\leq c \left(\int_0^t \|\nabla s(\tau)\|_2^2 \, d\tau \right)^{1/q'} \left(\int_0^t \|s(\tau)\|_2^2 \|s_1(\tau)\|_p^q \, d\tau \right)^{1/q}$$

$$\leq \nu \int_0^t \|\nabla s(\tau)\|_2^2 \, d\tau + c \int_0^t \|s(\tau)\|_2^2 \|s_1(\tau)\|_p^q \, d\tau, \quad (3.23)$$

by lemma 10.2.5 with an appropriate choice of η. Replacing the right side of (3.22) by (3.23), we obtain

$$\|s(t)\|_2^2 \leq c \int_0^t \|s(\tau)\|_2^2 \|s_1(\tau)\|_p^q \, d\tau. \quad (3.24)$$

Write

$$f(t) = \left\{ \exp \left[-c \int_0^t \|s_1(\tau)\|_p^q \, d\tau \right] \right\} \int_0^t \|s(\tau)\|_2^2 \|s_1(\tau)\|_p^q \, d\tau.$$

It is easy to see that (3.24) is equivalent to

$$\frac{d}{dt} f(t) \leq 0.$$

Therefore, f is non-increasing. Since f is initially zero and can never be negative, this implies $f(t) \equiv 0$. Therefore,

$$\int_0^t \|s(\tau)\|_2^2 \, \|s_1(\tau)\|_p^q \, d\tau \equiv 0.$$

Returning now to (3.24), we conclude that $s(t) \equiv 0$, and this is the same as $s_2(t) \equiv s_1(t)$.

The energy inequalities (3.20) and (3.21) were used in this proof. But on any interval $[t_0, t_0 + T)$ for which $t_0 \in S_0$, the set defined in theorem 11.5.5, similar inequalities are valid. Therefore, we have,

COROLLARY 3.4 *Let V be bounded. Let s_1 and s_2 be weak solutions of the Navier-Stokes equations with $s_1(t_0) = s_2(t_0)$, where t_0 lies in the set S_0 of theorem 11.5.5. If $s_1 \in L^q((t_0, t_0 + T); L^p)$, where $3/p + 2/q = 1$, then $s_2(t) = s_1(t)$ for every $t \in [t_0, t_0 + T)$.*

In § 11.5, we proved that weak solutions satisfy only an energy *inequality*, rather than the energy equality that was derived in § 7.4 for classical solutions. We can now say when a weak solution has the stronger property derived in § 7.4. When a weak solution $s_1 \in L^q((0, T); L^p)$, (3.14) holds for any other weak solution s_2 with the same initial values as s_1. Set $s_2 = s_1$ in (3.14). The result is

$$\|s_1(t)\|_2^2 + 2\nu \int_0^t \|\nabla s_1(\tau)\|_2^2 \, d\tau = \|s^0\|_2^2, \tag{3.25}$$

which is precisely what one gets from the energy equation (7.4.3) by integration. Thus, we have prove the first part of

THEOREM 3.4 *Let s_1 be a weak solution of the Navier-Stokes equations satisfying $s_1 \in L^p((0, T); L^p)$, where*

$$\frac{3}{p} + \frac{2}{q} = 1.$$

Then, s_1 satisfies the integrated energy equation (3.25) for $t \in [0, T)$. Consequently, s_1 is continuous (as a function of t) in the strong topology of $\mathscr{L}^2$.

We leave the proof of the last part of the theorem as an exercise.

Exercises

3.1   Prove the last sentence of theorem 3.4.

3.2   Prove that in two dimensions, theorem 3.3 becomes the following. *Let s_1 and s_2 be weak solutions of the Navier-Stokes equations in two dimensions, with $s_1(0) = s_2(0)$. If $s_1 \in L^q((0, T); L^p)$, where*

$$\frac{2}{p} + \frac{2}{q} = 1,$$

then $s_2(t) = s_1(t)$ for every $t \in [0, T)$. Then, use the energy inequality to prove that *the weak solution is unique in two dimensions.*

CHAPTER 13

Strong solutions

1 Introduction

THE WEAK solution of the last two chapters is in some sense a solution of the Navier–Stokes equations, although it is not clear that the derivatives $\dot{s}$ and $\nabla^2 s$ appearing in those equations ever have any meaning. In this chapter, we consider solutions for which these derivatives exist as elements of $L^2(I; L^2)$ for certain intervals I. Such solutions are called strong. For strong solutions, each term in the Navier–Stokes equations exists as an element of the Hilbert space $L^2(I; L^2)$, at least. Actually, strong solutions are infinitely differentiable in all variables and all the terms in the equations are continuous, but this is not proved here[1].

To be specific, *a strong solution* of the Navier-Stokes equations on the interval $[0, T)$ is a weak solution that is in one of the spaces $\mathscr{K}(a, T)$ defined in chapter ten. Recall that $\mathscr{K}(a, T)$ is the completion of $C_0^1([0, T); \mathscr{H}^2)$ with respect to the norm

$$\|\varphi\|_{\mathscr{K}(a,T)} = \left[\|\nabla\varphi(0)\|_2^2 + \int_0^T e^{2at} (\|\dot{\varphi}(t)\|_2^2 + \|P\nabla^2\varphi(t)\|_2^2)\, dt \right]^{1/2}. \quad (1.1)$$

If φ is any element of $\mathscr{K}(a, T)$, the operators d/dt and $P\nabla^2$ make sense when they act on φ. To see this, let $\varphi \in \mathscr{K}(a, T)$, and let $\{\varphi_n\}$ be a sequence in $C_0^1([0, T); \mathscr{H}^2)$ converging to φ in $\mathscr{K}(a, T)$. Since $\{\varphi_n\}$ converges in $\mathscr{K}(a, T)$, it is Cauchy there, and (1.1) shows that the sequences $\{\dot{\varphi}_n\}$ and $\{P\nabla^2\varphi_n\}$ are Cauchy in $L^2((0, T); L^2)$. Therefore, these sequences converge in $L^2((0, T); L^2)$, and we can define

$$\dot{\varphi} = \lim_{n \to \infty} \dot{\varphi}_n \quad (1.2)$$

and

$$P\nabla^2\varphi = \lim_{n \to \infty} P\nabla^2\varphi_n, \quad (1.3)$$

the limits being in $L^2((0, T); L^2)$. It is easy to show that these definitions are independent of the particular sequence $\{\varphi_n\}$ chosen to represent φ.

We have occasion later on to talk about strong solutions on intervals not of the form $[0, T)$. If I is any interval $[t_0, t_1)$, we say that s is a *strong solution*

---

[1] See Shmuel Kaniel and Marvin Shinbrot, Smoothness of weak solutions of the Navier-Stokes equations. Archive for rational mechanics and analysis, 24 (1967) 302–324. Some errors made in the reference are corrected here.

on I if the function s_{t_0} defined by $s_{t_0}(t) = s(t - t_0)$ is an element of $\mathscr{K}(a, t_1 - t_0)$ and s is a weak solution on $[t_0, t_1)$.

The Navier–Stokes equations are, of course,

$$\dot{s}(t) + s(t) \cdot \nabla s(t) - \nu \nabla^2 s(t) = -\frac{1}{\varrho} \nabla \pi(t), \qquad (1.4)$$

$$\nabla \cdot s(t) = 0, \qquad (1.5)$$

if the external forces are zero. These equations are to be solved along with the boundary condition

$$s(t) = 0 \quad \text{on } \partial V \qquad (1.6)$$

and the initial condition

$$s(0) = s^0. \qquad (1.7)$$

As we saw in chapter eleven, a weak solution exists whenever $s^0 \in \mathscr{L}^2$. In this chapter, we consider initial data in a slightly better space, namely, $\mathscr{H}_0^{1,2}$. We prove three principal results. First, we show that if $s^0 \in \mathscr{H}_0^{1,2}$, then there is a strong solution in $\mathscr{K}(a, T)$ if T is small enough. Then, we show that if the initial data are small enough (in the norm of $\mathscr{H}_0^{1,2}$), there is a strong solution in $\mathscr{K}(a, \infty)$ for some $a > 0$. Third, we examine the set of values of t on which a weak solution is *not* strong, and we show that this set must be rather small.

It is perhaps worth another mention that the parameter a occurring in the definition (1.1) is irrelevant if $T < \infty$, since the norms in $K(a_1, T)$ and $K(a_2, T)$ are equivalent for any a_1 and a_2 when $T < \infty$. On the other hand, when $T = \infty$, to say that a weak solution is strong is to say that it is exponentially small for large t. Thus, the statement that a weak solution is in $\mathscr{K}(a, \infty)$ is a kind of stability theorem.

2 Some preliminaries

We prove here a few results needed in our later discussions of strong solutions. The first is a variant of the Sobolev inequality (10.2.6) with $p = 2$. It is a lemma on functions in $\mathscr{H}^2$ (see p. 148). To see its significance for strong solutions, notice that (1.1) shows that if $\varphi \in \mathscr{K}(a, T)$, then $\varphi(t) \in \mathscr{H}^2$ for almost all $t \in [0, T)$.

LEMMA 2.1 $\mathscr{H}^2 \subset \mathscr{H}_0^{1,6}$. *Moreover, if* $\varphi \in \mathscr{H}^2$, *then*

$$\|\nabla \varphi\|_6 \leqslant c \|P\nabla^2 \varphi\|_2. \qquad (2.1)$$

Proof Let φ and ψ be elements of $\mathscr{C}_*^\infty$ and $\mathscr{C}_0^\infty$, respectively. Then, $\psi \in \mathscr{L}^2$, so that

$$|(\nabla \varphi, \nabla \psi)| = |(\nabla^2 \varphi, \psi)| = |(P\nabla^2 \varphi, \psi)|$$

$$\leqslant \|P\nabla \varphi^2\|_2 \|\psi\|_2 \leqslant c \|P\nabla^2 \varphi\|_2 \|\nabla \psi\|_{6/5},$$

by lemma 10.2.6. Therefore, if $\psi \neq 0$,

$$\frac{|(\nabla\varphi, \nabla\psi)|}{\|\nabla\psi\|_{6/5}} \leqslant c \, \|P\nabla^2\varphi\|_2.$$

Now, the dual of $\mathscr{H}_0^{1,6}$ is $\mathscr{H}_0^{1,6/5}$ (exercise 10.2.3). Therefore, the supremum of the left hand side over all $\psi \in \mathscr{C}_0^\infty$ is just the norm of φ in $\mathscr{H}_0^{1,6}$. This proves (2.1) for $\varphi \in \mathscr{C}_*^\infty$. Since $\mathscr{C}_*^\infty$ is dense in $\mathscr{H}^2$, the lemma follows.

LEMMA 2.2 $\mathscr{H}^2 \subset \mathscr{L}^2$, and if V is bounded, the injection of $\mathscr{H}^2$ into $\mathscr{L}^2$ is compact.

Proof That $\mathscr{H}^2 \subset \mathscr{L}^2$ is clear from lemma 2.1 and lemma 10.2.7. To prove the compactness, we have to show that if $\{\varphi_k\}$ is a weakly convergent sequence in $\mathscr{H}^2$, then $\{\varphi_k\}$ converges strongly in $\mathscr{L}^2$. Let $\{\varphi_k\}$ converge weakly to φ in $\mathscr{H}^2$. Lemmas 2.1 and 10.2.7 show that $\{\varphi_k\}$ converges weakly to φ in $\mathscr{L}^2$, while lemmas 2.1 and 10.2.3 show that

$$\|\nabla\varphi_k\|_2 \leqq c \, \|P\nabla^2\varphi_k\|_2,$$

and this is bounded. Let $\{\omega_n\}$ be the sequence of lemma 10.2.9. By that lemma,

$$\|\varphi_k - \varphi\|_2^2 \leqslant \sum_1^N |(\varphi_k - \varphi, \omega_n)|^2 + \varepsilon \, \|\nabla\varphi_k - \nabla\varphi\|_2^2,$$

$$\leqslant \sum_1^N |(\varphi_k - \varphi, P\omega_n)|^2 + c\varepsilon.$$

The second term can be made as small as desired by choosing ε small enough. Having chosen ε, N is fixed, and we can make the first term as small as we please by choosing k large enough, since $\{\varphi_k\}$ converges weakly to φ in $\mathscr{L}^2$. Therefore, $\{\varphi_k\}$ converges strongly to φ in $\mathscr{L}^2$, and lemma 2.2 is proved.

LEMMA 2.3 *There exists a sequence $\{\sigma_n\} \subset \mathscr{L}^2$, complete and orthogonal in $\mathscr{L}^2$, satisfying*

$$-P\nabla^2\sigma_n = \lambda_n\sigma_n,$$

where λ_n is real for each n and $\lambda_n \to \infty$.

Proof We define an operator A on $\mathscr{L}^2$ as follows. $\varphi = A\psi$ if $\varphi \in \mathscr{H}^2$ and $-P\nabla^2\varphi = \psi$. A is well-defined since, as we saw in chapter ten, $\|P\nabla^2\varphi\|_2$ being zero implies φ is zero. A is also bounded, for let $\varphi \in \mathscr{H}^2$. Then, lemmas 2.1 and 10.2.3 show that

$$\|\varphi\|_2 \leqq c \, \|P\nabla^2\varphi\|_2.$$

Writing $\varphi = A\psi$, we find that this means

$$\|A\psi\|_2 \leqq c \, \|\psi\|_2.$$

Next, the operator A is compact, since whenever $\{\psi_n\}$ converges weakly to ψ in $\mathscr{L}^2$, lemma 2.2 shows that $\{A\psi_n\}$ converges strongly to $A\psi$. Also, A is self-adjoint, for if $\varphi_1 = A\psi_1$ and $\varphi_2 = A\psi_2$, we have, using Green's theorem in the second step,

$$(A\psi_1, \psi_2) = -(\varphi_1, P\nabla^2 \varphi_2) = -(P\nabla^2\varphi_1, \varphi_2) = (\psi_1, A\psi_2)$$

since φ_1 and φ_2 are in $\mathscr{L}^2$. Finally the nullspace of A is zero. Therefore, A has an associated sequence of eigenfunctions $\{\sigma_n\}$, complete and ortho-normal in $\mathscr{L}^2$. If the eigenvalue corresponding to σ_n is $1/\lambda_n$, then $\lambda_n \to \infty$, and

$$\lambda_n \sigma_n = -P\nabla^2 \sigma_n.$$

This proves lemma 2.3.

The next result tells us that any strong solution of (1.4–7) satisfies the boundary condition (1.6) automatically.

LEMMA 2.4   *Let $\varphi \in K(a, T)$. If ∂V is smooth enough, then for almost all $t \in [0, T), \varphi(t)$ is zero on ∂V.*

Proof   Lemma 2.1 shows that if $\varphi \in \mathscr{K}(a, T)$, then $\varphi(t) \in \mathscr{H}_0^{1,6}$ almost everywhere. Lemma 10.2.8 then shows that $\varphi(t) \in C^0(\bar{V})$, and exercise 10.2.1 then gives the result.

Another fact about the space $K(a, T)$ is expressed in

LEMMA 2.5   *Let $\varphi \in \mathscr{K}(a, T)$. Then φ can be redefined on a set of values of t having Lebesgue measure zero in such a way that $\|\nabla\varphi(t)\|^2$ becomes an abso-lutely continuous function of t, and $(d/dt)\|\nabla\varphi(t)\|_2^2 = -2\,(\dot{\varphi}(t), P\nabla^2\varphi(t))$. Consequently, $\varphi(t) \in \mathscr{H}_0^{1,2}$ for all $t \in [0, T)$.*

Proof   By (1.1), (1.2) and (1.3), there is a sequence $\{\varphi_n\} \subset C_0^1([0, T); \mathscr{H}^2)$ such that

$$\varphi_n(0) \to \varphi(0) \quad \text{in} \quad \mathscr{H}_0^{1,2},$$

$$\dot{\varphi}_n \to \dot{\varphi} \quad \text{in} \quad L^2((0, T); L^2)$$

$$P\nabla^2\varphi_n \to P\nabla^2\varphi \quad \text{in} \quad L^2((0, T); \mathscr{L}^2). \tag{2.2}$$

Also, $\dot{\varphi}_n$ takes values in $\mathscr{L}^2$.

Let $t < T$, and consider

$$\int_0^t (\dot{\varphi}(\tau),\, P\nabla^2\varphi\,(\tau))\, d\tau = \lim \int_0^t (\dot{\varphi}_n(\tau),\, P\nabla^2\varphi_n\,(\tau))\, d\tau$$

$$= \lim \int_0^t (\dot{\varphi}_n(\tau),\, \nabla^2\varphi_n\,(\tau))\, d\tau$$

$$= -\lim \int_0^t (\nabla\dot{\varphi}_n\,(\tau),\, \nabla\varphi_n\,(\tau))\, d\tau$$

$$= -\frac{1}{2} \lim \int_0^t \frac{d}{dt} \|\nabla\varphi_n\,(\tau)\|_2^2\, d\tau$$

$$= \frac{1}{2} \|\nabla\varphi\,(0)\|_2^2 - \frac{1}{2} \lim \|\nabla\varphi_n\,(t)\|_2^2;$$

Green's theorem is used in the third step. (2.2) shows that $\lim \|\nabla\varphi_n(t)\|_2^2$ exists almost everywhere. (2.2) and lemmas 2.1 and 10.2.3 show that this limit must be $\|\nabla\varphi(t)\|_2^2$ almost everywhere. Upon redefinition on a set of measure zero, then, we find

$$\int_0^t (\dot{\varphi}(\tau),\, P\nabla^2\varphi\,(\tau))\, d\tau = \tfrac{1}{2} \|\nabla\varphi\,(0)\|_2^2 - \tfrac{1}{2} \|\nabla\varphi\,(t)\|_2^2. \tag{2.3}$$

Lemma 2.5 follows from this.

In all that follows, it is assumed that whenever $\varphi \in K(a, T)$, the adjustment on a set of measure zero has been made so that (2.3) holds for all $t \in [0, T)$.

Next, we prove

LEMMA 2.6 *If a is small enough, a norm equivalent to the norm on $K(a, T)$ is*

$$\left[\nu \|\nabla\varphi\,(0)\|_2^2 + \int_0^T e^{2at} \|\dot{\varphi}(t) - \nu P\nabla^2\varphi\,(t)\|_2^2\, dt\right]^{1/2}.$$

Proof Let $\varphi \in C_0^1\,([0, T); \mathscr{H}^2)$. Lemma 2.5 shows that

$$e^{2at} \|\dot{\varphi}(t) - \nu P\nabla^2\varphi\,(t)\|_2^2$$

$$= e^{2at} [\|\dot{\varphi}(t)\|_2^2 - 2\nu (\dot{\varphi}(t),\, P\nabla^2\varphi\,(t)) + \nu^2 \|P\nabla^2\varphi\,(t)\|_2^2] \tag{2.4}$$

$$= \nu \frac{d}{dt} (e^{2at} \|\nabla\varphi\,(t)\|_2^2) + e^{2at} [\|\dot{\varphi}(t)\|_2^2 + \nu^2 \|P\nabla^2\varphi\,(t)\|_2^2$$

$$- 2a\nu \|\nabla\varphi\,(t)\|_2^2].$$

Integrate this equation from 0 to T. Since φ has compact support in $[0, T)$, we find

$$\nu \|\nabla\varphi\,(0)\|_2^2 + \|e^{at}\,[\dot{\varphi}(t) - \nu P\nabla^2\varphi\,(t)]\|_{2,2}^2$$

$$= \|e^{at}\dot{\varphi}\,(t)\|_{2,2}^2 + \nu^2 \,\|e^{at}P\nabla^2\varphi\,(t)\|_{2,2}^2 - 2a\nu\,\|e^{at}\nabla\varphi\,(t)\|_{2,2}^2$$

$$\geqslant \|e^{at}\dot{\varphi}\,(t)\|_{2,2}^2 + \nu\,(\nu - 2ac)\,\|e^{at}P\nabla^2\varphi\,(t)\|_{2,2}^2$$

by lemma 2.1. If a is small enough, then $\nu - 2ac > 0$, and

$$2\,\{\nu\,\|\nabla\varphi\,(0)\|_2^2 + \|e^{at}\,[\dot{\varphi}(t) - \nu P\nabla^2\varphi\,(t)]\|_{2,2}^2\}$$

$$\geqslant 2\nu\,\|\nabla\varphi\,(0)\|_2^2 + \|e^{at}\,[\dot{\varphi}(t) - \nu P\nabla^2\varphi\,(t)]\|_{2,2}^2$$

$$\geqslant \nu\,(\nu - 2ac)\,[\|\nabla\varphi\,(0)\|_2^2 + \|e^{at}\dot{\varphi}\,(t)\|_{2,2}^2 + \|e^{at}P\nabla^2\varphi\,(t)\|_{2,2}^2]$$

$$= \nu\,(\nu - 2ac)\,\|\varphi\|_{K(a,T)}^2.$$

To prove an inequality in the opposite direction, we return to (2.4). We have

$$e^{2at}\,\|\dot{\varphi}(t) - \nu P\nabla^2\varphi\,(t)\|_2^2$$

$$\leqslant e^{2at}\,[\|\dot{\varphi}(t)\|_2^2 + 2\nu\,\|\dot{\varphi}(t)\|_2\,\|P\nabla^2\varphi\,(t)\|_2 + \nu^2\,\|P\nabla^2\varphi\,(t)\|_2^2]$$

$$\leqslant ce^{2at}\,[\|\dot{\varphi}(t)\|_2^2 + \nu^2\,\|P\nabla^2\varphi\,(t)\|_2^2].$$

Integrating and adding $\nu\,\|\nabla\varphi(0)\|_2^2$, we find

$$\nu\,\|\nabla\varphi(0)\|_2^2 + \|e^{at}\,[\dot{\varphi}(t) - \nu P\nabla^2\varphi\,(t)]\|_{2,2}^2 \leqslant c\,\|\varphi\|_{K(a,T)}^2.$$

This proves lemma 2.6.

LEMMA 2.7   *Let* $\varphi \in \mathscr{K}\,(a,T)$. *Then*, $\varphi(t)\cdot\nabla\varphi(t) \in L^2$ *for almost all* $t \in (0,T)$.

Proof   Hölder's inequality gives

$$\|\varphi(t)\cdot\nabla\varphi\,(t)\|_2 \leq \|\varphi(t)\|_\infty\,\|\nabla\varphi\,(t)\|_2$$

$$\leq c\,\|\nabla\varphi\,(t)\|_6\,\|\nabla\varphi\,(t)\|_1,$$

by lemma 10.2.8. Lemma 2.1 now gives

$$\|\varphi(t)\cdot\nabla\varphi\,(t)\|_2 \leq c\,\|P\nabla^2\varphi\,(t)\|_2\,\|\nabla\varphi\,(t)\|_2.$$

But lemma 2.5 shows hat

$$\|\nabla\varphi\,(t)\|_2^2 = \|\nabla\varphi\,(0)\|_2^2 - 2\int_0^t (\dot{\varphi}(\tau),\,P\nabla^2\varphi\,(\tau))\,d\tau$$

$$\leq c\,\|\varphi\|_{\mathscr{K}(a,T)}^2$$

if $0 \leq t < T$. Therefore,

$$\|\varphi(t)\cdot\nabla\varphi\,(t)\|_2 \leq c\,\|\varphi\|_{\mathscr{K}(a,T)}\,\|P\nabla^2\varphi\,(t)\|_2,$$

and this is almost everywhere finite since $\varphi(t) \in \mathscr{H}^2$ almost everywhere.

Next, we prove

LEMMA 2.8 *Let* $\varphi \in \mathscr{K}\ (a, T)$. *Then,* $\dot{\varphi}(t) \in \mathscr{L}^2$ *a.e. in* $(0, T)$.

Proof That $\dot{\varphi}(t) \in L^2$ a.e. is clear from the definition of $\mathscr{K}\ (a, T)$. The problem is only to show that $\varphi(t)$ lies in the subspace $\mathscr{L}^2$. For this purpose, let $\psi \in (\mathscr{L}^2)^\perp$. Then, $(\dot{\varphi}(t), \psi) = (d/dt)\,(\varphi(t), \psi)$. But $\varphi(t) \in \mathscr{H}^2$, while $\mathscr{H}^2 \subset \mathscr{L}^2$. Therefore, $(\varphi(t), \psi) = 0$, and, therefore, $(\dot{\varphi}(t), \psi) = 0$ a.e.

3 Uniqueness of strong solutions

It will be important in § 5 to know when strong solutions are unique. To discuss that question, we need

LEMMA 3.1 *Let* s *be a strong solution of the Navier-Stokes equations in an interval* $[t_0, t_0 + T)$. *Then,* s *satisfies the energy equation*

$$\tfrac{1}{2}\|s(t)\|_2^2 + \nu \int_{t_0}^t \|\nabla s\,(\tau)\|_2^2\,d\tau = \tfrac{1}{2}\|s(t_0)\|_2^2,\ t_0 \leqslant t < t_0 + T. \tag{3.1}$$

(3.1) is obtained, of course, by multiplying (1.4) by $s(t)$ and integrating. Strong solutions have enough smoothness to allow the computations of § 7.4 to be made.

Let s be a strong solution on $[t_0, t_0 + T)$. (1.1) and lemma 2.1 show that $s \in L^2\ ((t_0, t_0 + T);\ \mathscr{H}_0^{1,6})$. Lemma 10.2.8 then shows that $s \in L^2\ ((t_0, t_0 + T);\ C^0(\bar{V})) \subset L^2\ ((t_0, t_0 + T);\ L^\infty)$. Corollary 12.3.4 then implies

THEOREM 3.2 *Let* σ *be a weak solution of the Navier-Stokes equations, and let* t_0 *be a point in the set* S_0 *defined in theorem 11.5.5. Let* s *be a strong solution in an interval* $[t_0, t_0 + T)$. *If* $\sigma(t_0) = s(t_0)$, *then* $\sigma(t) = s(t)$ *throughout* $[t_0, t_0 + T)$.

4 Some a priori inequalities

Strong solutions were first proved to exist by A. A. Kiselev and O. A. Lady-zhenskaya[1], who assumed the initial data s^0 to be in $\mathscr{H}^2 \cap \mathscr{H}_0^{1,2}$. We prove a stronger result in this section, that a strong solution exists if $s^0 \in \mathscr{H}_0^{1,2}$. The argument in this case is no more difficult, and we prove in the next section that this apparently small change in the space in which the initial data lie

---

[1] On the existence and uniqueness of the solution of the nonstationary problem for a viscous incompressible fluid. Izvestia Akad. Nauk USSR, ser. mat. 21 (1957) 655–680. A more accessible reference for those who do not read Russian is the book by Ladyzhenskaya cited on p. 111.

makes a big difference in some of the conclusions that can be drawn. That there is a strong solution if $s^0 \in \mathscr{H}_0^{1,2}$ was first proved by Giovanni Prodi[1] and, independently, by Shmuel Kaniel and me[2].

We saw in chapter eleven that it was certain inequalities that supplied the existence of a weak solution. Such inequalities, valid if a solution exists, and implying its existence, are called *a priori* inequalities. The strong solution is no different from the weak solution, in that it satisfies an *a priori* inequality. The inequality it satisfies is stated in

THEOREM 4.1   *Let V be bounded. If $s^0 \in \mathscr{H}_0^{1,2}$, then any strong solution of* (1.4–7) *satisfies*

$$\|s\|_{\mathscr{X}(a,T)} \leq \frac{c \, \|\nabla s^0\|_2}{(1 - cT \, \|\nabla s^0\|_2^4)^{1/4}} \qquad (4.1)$$

whenever T is small enough that the right side of (4.1) *makes sense.*

Proof   If s is a strong solution, lemma 2.8 shows that $s(t) \cdot \nabla s(t) \in L^2$ a.e., while lemma 2.9 shows that $\dot{s}(t) \in \mathscr{L}^2$ a.e. Therefore, operating on (1.4) with P, we find

$$\dot{s}(t) + P\,[s(t) \cdot \nabla s\,(t)] - \nu P\nabla^2 s\,(t) = 0. \qquad (4.2)$$

Multiply (4.2) by $-P\nabla^2 s\,(t)$ and integrate over V. Lemma 2.6 shows that the result is

$$\frac{1}{2}\frac{d}{dt}\,\|\nabla s\,(t)\|_2^2 + \nu\,\|P\nabla^2 s\,(t)\|_2^2$$

$$= (\dot{s}(t) \cdot \nabla s\,(t), P\nabla^2 s\,(t)) \leq \|s(t) \cdot \nabla s\,(t)\|_2 \,\|P\nabla^2 s\,(t)\|_2. \qquad (4.3)$$

Using Hölder's inequality, lemma 10.2.6, lemma 10.2.4, and then lemma 2.1, we find

$$\|s(t) \cdot \nabla s\,(t)\|_2 \leq \|s(t)\|_{12}\,\|\nabla s\,(t)\|_{12/5} \leq c\,\|\nabla s\,(t)\|_{12/5}^2$$

$$\leq c\,\|\nabla s\,(t)\|_2^{3/2}\,\|\nabla s\,(t)\|_6^{1/2} \leq c\,\|\nabla s\,(t)\|_2^{3/2}\,\|P\nabla^2 s\,(t)\|_2^{1/2}. \qquad (4.4)$$

Substituting this into (4.3) and then using lemma 10.2.5, we find, choosing η

[1] Giovanni Prodi, Teoremi di tipo locale per il sistema di Navier–Stokes e stabilità della soluzioni stazionarie. Rend. Sem. Mat. Univ. Padova (1962) 374–397.

[2] The initial value problem for the Navier–Stokes equations. Archive for rational mechanics and analysis, 21 (1966) 270–285.

appropriately,

$$\frac{1}{2} \frac{d}{dt} \|\nabla s(t)\|_2^2 + \nu \|P\nabla^2 s(t)\|_2^2 \leqslant c\|\nabla s(t)\|_2^{3/2} \|P\nabla^2 s(t)\|_2^{3/2}$$

$$\leqslant c \|\nabla s(t)\|_2^6 + \frac{\nu}{2} \|P\nabla^2 s(t)\|_2^2. \qquad (4.5)$$

Therefore,

$$\frac{d}{dt} \|\nabla s(t)\|_2^2 + \nu \|P\nabla^2 s(t)\|_2^2 \leqslant c \|\nabla s(t)\|_2^6. \qquad (4.6)$$

Throwing away the second term in (4.6), we find that it becomes

$$\frac{d}{dt} \|\nabla s(t)\|_2^{-4} \geqslant -c.$$

This inequality can be integrated to give

$$\frac{1}{\|\nabla s(t)\|_2^4} \geqslant \frac{1}{\|\nabla s^0\|_2^4} - ct.$$

As long as the right hand side is positive, then,

$$\|\nabla s(t)\|_2 \leqslant \frac{\|\nabla s^0\|_2}{(1 - ct \|\nabla s^0\|_2^4)^{1/4}}, \qquad (4.7)$$

and (4.6) gives

$$\frac{d}{dt} \|\nabla s(t)\|_2^2 + \nu \|P\nabla^2 s(t)\|_2^2 \leqslant \frac{c \|\nabla s^0\|_2^6}{(1 - ct \|\nabla s^0\|_2^4)^{3/2}}. \qquad (4.8)$$

Consider next $\|\dot s(t) - \nu P\nabla^2 s(t)\|_2^2$. By lemma 2.5,

$$\|\dot s(t) - \nu P\nabla^2 s(t)\|_2^2 = \|\dot s(t)\|_2^2 + \nu \frac{d}{dt} \|\nabla s(t)\|_2^2 + \nu^2 \|P\nabla^2 s(t)\|_2^2.$$

On the other hand, by (4.2),

$$\|\dot s(t) - \nu P\nabla^2 s(t)\|_2^2 = \|P[s(t) \cdot \nabla s(t)]\|_2^2$$

$$\leqslant \|s(t) \cdot \nabla s(t)\|_2^2$$

$$\leqslant c \|\nabla s(t)\|_2^3 \|P\nabla^2 s(t)\|_2$$

by (4.4). Therefore,

$$\|\dot s(t)\|_2^2 + \nu \frac{d}{dt} \|\nabla s(t)\|_2^2 + \nu^2 \|P\nabla^2 s(t)\|_2^2 \leqslant c \|\nabla s(t)\|_2^6 + \nu^2 \|P\nabla^2 s(t)\|_2^2,$$

again by lemma 10.2.5. This inequality and (4.7) give

$$\|\dot{s}\|_2^2 + v \frac{d}{dt} \|\nabla s(t)\|_2^2 \leqslant c \frac{\|\nabla s^0\|_2^6}{(1 - ct \|\nabla s^0\|_2^4)^{3/2}}$$

and, integrating,

$$\int_0^T \|\dot{s}(t)\|_2^2 \, dt \leqslant v \|\nabla s^0\|_2^2 + c \|\nabla s^0\|_2^2 \left[\frac{1}{(1 - ct \|\nabla s^0\|_2^4)^{1/2}} - 1 \right]$$

$$\leqslant \frac{c \|\nabla s^0\|_2^2}{(1 - cT \|\nabla s^0\|_2^4)^{1/2}}.$$

Integrating (4.8), we find that $\int_0^T \|P\nabla^2 s(t)\|_2^2 \, dt$ satisfies a similar inequality. Therefore,

$$\|s\|_{K(0,T)}^2 = \|\nabla s^0\|_2^2 + \int_0^T [\|\dot{s}(t)\|_2^2 + \|P\nabla^2 s(t)\|_2^2] \, dt$$

$$\leqslant \frac{c \|\nabla s^0\|_2^2}{(1 - cT \|\nabla s^0\|_2^4)^{1/2}}.$$

This is equivalent to (4.1) with $a = 0$. Since (4.1) only makes sense for $T < \infty$, and since the norms in $K(a, T)$ and $K(0, T)$ are equivalent for $T < \infty$, we have proved theorem 4.1.

Another *a priori* inequality is proved in

THEOREM 4.2 *Let V be bounded. If $s^0 \in \mathscr{H}_0^{1,2}$ and $\|\nabla s^0\|_2$ is small enough, then any strong solution of (1.4–7) satisfies*

$$\|s\|_{K(a,\infty)} \leqslant c \|\nabla s^0\|_2. \tag{4.9}$$

The value of a for which (4.9) holds depends on $\|\nabla s^0\|_2$ and is proportional to $\|\nabla s^0\|_2^4$. It is also true that

$$\|s\|_{K(a,\infty)} \leqslant c \|\nabla s^0\|_2 (1 + \|\nabla s^0\|_2^4)^{1/2} \tag{4.10}$$

for some sufficiently small value of a independent of $\|\nabla s^0\|^2$.

To prove (4.9), we begin with (4.5). We have

$$\frac{d}{dt} \|\nabla s(t)\|_2^2 \leqslant c \|\nabla s(t)\|_2^6 - v \|P\nabla^2 s(t)\|_2^2$$

$$\leqslant (c \|\nabla s(t)\|_4^2 - v) \|\nabla s(t)\|_2^2, \tag{4.11}$$

by lemma 2.1. If $\|\nabla s^0\|_2$ is small enough, the right side of (4.11) is negative when $t = 0$. Therefore, $\|\nabla s(t)\|_2$ is initially decreasing. Since decreasing $\|\nabla s(t)\|_2$ only has the effect of making the right side of (4.11) more strongly negative, we infer that $\|\nabla s(t)\|_2$ always decreases. Therefore,

$$\|\nabla s(t)\|_2 \leqslant \|\nabla s^0\|_2. \tag{4.12}$$

But we can say even more. If $\|\nabla s^0\|_2$ is small enough,

$$c \, \|\nabla s^0\|_2^4 - \nu \leqslant -\frac{\nu}{2} \qquad (4.13)$$

and, since $\|\nabla s(t)\|_2$ is decreasing, if (4.13) is true, it is also true with s^0 replaced by $s(t)$. Therefore,

$$c \, \|\nabla s(t)\|_2^4 - \nu \leqslant -\frac{\nu}{2}.$$

Inserting this on the right side of (4.11), we find

$$\frac{d}{dt} \|\nabla s(t)\|_2^2 \leqslant -\frac{\nu}{2} \|\nabla s(t)\|_2^2 \qquad (4.14)$$

$$\leqslant -c \, \|\nabla s^0\|_2^4 \, \|\nabla s(t)\|_2^2 \qquad (4.15)$$

(4.15) is equivalent to

$$\frac{d}{dt} \, [e^{ct\|\nabla s^0\|_2^4} \, \|\nabla s(t)\|_2^2] \leqslant 0.$$

Therefore the quantity in brackets is non-increasing, and we conclude

$$\|\nabla s(t)\|_2^2 \leqslant \|\nabla s^0\|_2^2 \, e^{-ct\|\nabla s^0\|_2^4}. \qquad (4.16)$$

As in the proof of theorem 4.1, we have

$$\|\dot{s}(t) - \nu P \nabla^2 s(t)\|_2^2 \leqslant c \, \|\nabla s(t)\|_2^6. \qquad (4.17)$$

Let $2a = c_1 \|\nabla s^0\|_2^4$. Multiplying (4.17) by e^{2at} and integrating, we find, if c_1 is small enough,

$$\int_0^T \|e^{at} \, (\dot{s}(t) - \nu P \nabla^2 s(t))\|_2^2 \, dt \leqslant c \, \|\nabla s^0\|_2^6 \int_0^\infty e^{-c_2 t\|\nabla s^0\|_2^4} \, dt = c \, \|\nabla s^0\|_2^2.$$

Adding $\|\nabla s^0\|_2^2$ to both sides of this inequality and using lemma 2.6, we obtain (4.9).

To prove (4.10), we begin with (4.14) instead of (4.15) and prove

$$\|\nabla s(t)\|_2^2 \leqslant \|\nabla s^0\|_2^2 \, e^{-(\nu/4)t}$$

instead of (4.16). Then, again just as before, we find (4.10).

5 Existence of strong solutions

To prove there is a strong solution, we need a theorem on ordinary differential equations in a finite-dimensional space, at least part of which is well known.

LEMMA 5.1 *Let H be a finite-dimensional Hilbert space with norm $\|\cdot\|$. Let $F: H \to H$ be a function satisfying a Lipschitz condition of the form*

$$\|F(s^1) - F(s^2)\| \leq f(s^1, s^2) \|s^1 - s^2\|, \tag{5.1}$$

where f is a continuous, real valued function on $H \times H$. Let $t_0 \geq 0$. Then, there exist a $t_1 > t_0$ such that the initial value problem

$$\dot{s}(t) = F(s(t)), \tag{5.2}$$

$$s(t_0) = s^0, \tag{5.3}$$

has a solution in $C^1([t_0, t_1); H)$ for every $s^0 \in H$. If the solution is bounded in $[t_0, t_1)$, it can be continued beyond t_1.

All but the last sentence is the standard existence theorem on systems of ordinary differential equations, since H is finite dimensional. To prove the last sentence, let s be a solution of (5.2–3) that is bounded:

$$\|s(t)\| \leq R, \quad t_0 \leq t < t_1. \tag{5.4}$$

F is continuous, and H is finite dimensional. Therefore, F is bounded on the ball of radius R:

$$\|F(s)\| \leq c \quad \text{whenever} \quad \|s\| \leq R.$$

If s is a solution of (5.2–3), s clearly satisfies

$$s(t) = s^0 + \int_{t_0}^{t} F(s(\tau)) \, d\tau. \tag{5.5}$$

Let $t_0 \leq t' < t'' < t_1$. Then,

$$\|s(t'') - s(t')\| \leq \int_{t'}^{t''} \|F(s(\tau))\| \, d\tau \leq c(t'' - t'), \tag{5.6}$$

by (5.4) and (5.5). Since the right side of (5.6) goes to zero as t' and t'' increase to t_1, $\lim_{t \uparrow t_1} s(t)$ exists. Set

$$s^1 = \lim_{t \uparrow t_1} s(t).$$

Let s' be the solution of the problem

$$\dot{s}'(t) = F(s'(t))$$

$$s'(t_1) = s^1.$$

s' exists in $C^1([t_1, t_2); H)$ for some $t_2 > t_1$ by the first part of the lemma. Define

$$s''(t) = \begin{cases} s(t), & t_0 \leq t < t_1, \\ s'(t), & t_1 \leq t < t_2. \end{cases}$$

$s'' \in C^0 ([t_0, t_2); H)$, since $\lim_{t \uparrow t_1} s(t) = \lim_{t \downarrow t_1} s'(t)$. Moreover, $s'' \in C^1 ([t_0, t_2); H)$, since

$$\dot{s}''(t) = F(s''(t)),$$

while both F and s'' are continuous. Since s'' is clearly an extension of s to the larger interval $[t_0, t_2)$, and since s'' is a solution of (5.2–3) in $[t_0, t_2)$, the last part of the lemma is proved.

To return to the Navier – Stokes equations, we would like to solve

$$\dot{s}(t) + P [s(t) \cdot \nabla s(t)] - \nu P \nabla^2 s(t) = 0. \tag{5.7}$$

This promises to be difficult, so we project (5.9) down onto a finite-dimensional subspace of $\mathscr{L}^2$, solve the resulting equation there via lemma 5.1, and then try to climb back up to a solution of (5.7). This method of solving (5.7) is known as *Galerkin's method*.

In the following, the range of any linear operator L is denoted by $R(L)$.

LEMMA 5.2   *Let P_n be any orthogonal projection in L^2 with the following properties:*

$$R(P_n) \subset \mathscr{H}^2, \tag{5.8}$$

$$R(P_n) \quad \text{has dimension} \quad n < \infty. \tag{5.9}$$

Then, the initial value problem

$$\dot{s}(t) + P_n [s(t) \cdot \nabla s(t)] - \nu P_n \nabla^2 s(t) = 0 \tag{5.10}$$

$$s(0) = s_n^0 \tag{5.11}$$

has a solution in $C^1 ([0, \infty); R(P_n))$ for every s_n^0 in $R(P_n)$.

Proof   Notice that because of (5.8), $R(P_n) \subset \mathscr{L}^2$. Therefore, $P_n P = P_n$. Consequently, $P_n \nabla^2 s$ is defined for all $s \in R(P_n)$. Indeed, we may set

$$P_n \nabla^2 s = P_n (P \nabla^2 s),$$

while $P \nabla^2 s \in \mathscr{L}^2$ since $s \in \mathscr{H}^2$. Also, $s \cdot \nabla s \in L^2$ for all $s \in R(P_n)$ since, by an argument that should be familiar by now,

$$\|s \cdot \nabla s\|_2 \leqq c \, \|P \nabla^2 s\|_2^2$$

for all $s \in \mathscr{H}^2$. Therefore, we may set

$$F(s) = \nu P_n \nabla^2 s - P_n (s \cdot \nabla s),$$

and this function F is defined for all $s \in R(P_n)$.

F satisfies the conditions of lemma 5.1. To see this, let s^1 and s^2 be elements of $R(P_n)$. Let $\{\varphi_1, \varphi_2, \ldots, \varphi_n\}$ be a basis for $R(P_n)$. Then, s^1 and s^2 have the form

$$s^1 = \sum_{k=1}^n s_k^1 \varphi_k, \quad s^2 = \sum_{k=1}^n s_k^2 \varphi_k,$$

where the coefficients s_k^i are scalars. We have

$$F(s^1) - F(s^2)$$
$$= \nu \sum (s_k^1 - s_k^2) P_n \nabla^2 \varphi_k + \sum \sum [s_k^1 (s_j^1 - s_j^2) + (s_k^1 - s_k^2) s_j^2] P_n \varphi_k \cdot \nabla \varphi_j,$$

so that

$$\|F(s^1) - F(s^2)\|_2 \leq \nu \sum |s_k^1 - s_k^2| \, \|P_n \nabla^2 \varphi_k\|_2 + \sum \sum (|s_k^1| \, |s_j^1 - s_j^2|$$
$$+ |s_k^1 - s_k^2| \, |s_j^2|) \, \|P_n \varphi_k \cdot \nabla \varphi_j\|_2$$
$$\leq c_1 \left(\sum |s_k^1 - s_k^2|^2 \right)^{1/2} + c_2 \left(\sum [|s_k^1|^2 \right.$$
$$+ |s_k^2|^2] \big)^{1/2} \left(\sum |s_k^1 - s_k^2|^2 \right)^{1/2},$$

where

$$c_1^2 = \nu^2 \sum \|P_n \nabla^2 \varphi_k\|_2^2$$

and

$$c_2^2 = \sum \sum \|P_n \varphi_k \cdot \nabla \varphi_j\|_2^2.$$

The sequence $\{\varphi_1, \varphi_2, \ldots, \varphi_n\}$ can be chosen to be orthonormal in L^2. In that case,

$$\|s^1\|_2^2 = \sum |s_k^1|^2,$$

with similar formulas for $\|s^2\|_2^2$ and $\|s^1 - s^2\|_2^2$. Therefore,

$$\|F(s^1) - F(s^2)\|_2 \leq [c_1 + c_2 (\|s^1\|_2^2 + \|s^2\|_2^2)^{1/2}] \|s^1 - s^2\|_2.$$

This is the form (5.1) takes in this case.

Because of (5.9), lemma 5.1 shows that (5.10–11) has a solution in $C^1 ([0, t_1); R(P_n))$ for some $t_1 > 0$. We now show that this solution is defined for *all* $t \geq 0$. To do that, lemma 5.1 says that all we need do is show that the solution is bounded. Multiply (5.10) by $s(t)$ and integrate over V. Since Γ_n is an orthogonal projection and $s(t) \subset R(P_n)$ for $0 \leq t < t_1$, we obtain

$$(s(t), \dot{s}(t)) + (s(t) \cdot \nabla s(t), s(t)) - (\nabla^2 s(t), s(t)) = 0.$$

Since $s(t) \in R(P_n) \subset \mathscr{H}^2$, $s(t)$ is smooth enough for each term in this equation to be evaluated in the usual way, and we find

$$\frac{1}{2} \frac{d}{dt} \|s(t)\|_2^2 + \nu \|\nabla s(t)\|_2^2 = 0. \tag{5.12}$$

Of course, this is just the energy equation (7.4.3).

Integrating (5.12), we see that

$$\tfrac{1}{2} \|s(t)\|_2^2 + \nu \int_0^t \|\nabla s(\tau)\|_2^2 \, d\tau = \tfrac{1}{2} \|s_n^0\|_2^2, \tag{5.13}$$

which clearly entails

$$\|s(t)\|_2 \leqq \|s_n^0\|_2$$

for any t for which the solution exists. Lemma 5.1 now shows that the solution exists for all $t \geq 0$.

LEMMA 5.3   *Let V be bounded. If $s_n^0 \in \mathcal{H}_0^{1,2}$, the solution of (5.10–11) satisfies*

$$\|s\|_{\mathcal{X}(a,T)} \leqq \frac{c \, \|\nabla s_n^0\|_2}{(1 - cT \, \|\nabla s_n^0\|_2^4)^{1/4}}$$

whenever T is small enough for the right side of (5.14) to make sense. Also, if $\|\nabla s_n^0\|_2$ is small enough,

$$\|s\|_{\mathcal{X}(a,\infty)} \leqq c \, \|\nabla s_n^0\|_2 \, (1 + \|\nabla s_n^0\|_2^4)^{1/2}$$

for some sufficiently small value of a.

The proof of lemma 5.3 is virtually identical with the proofs of theorems 4.2 and 4.3. One multiplies (5.10) by $-P_n\nabla^2 s\,(t)$ and carries through the argument exactly as in theorems 4.2 and 4.3. The only change is that P is replaced by P_n whenever the former appears.

We can now prove that the Navier-Stokes equations have a strong solution.

THEOREM 5.4   *Let V be bounded. If $s^0 \in \mathcal{H}_0^{1,2}$, then the Navier – Stokes equations have a unique, strong solution in some interval $[0, T)$ with T small enough. If $\|\nabla s^0\|_2$ is small enough, the equations have a unique, strong solution in $[0, \infty)$.*

Proof   Let $\{\sigma_k\}$ be the sequence of eigenfunctions of the operator $-P\nabla^2$, derived in lemma 2.3. Since $\{\sigma_k\}$ is orthonormal in L^2, the operator P_n defined by

$$P_n s = \sum_{k=1}^{n} (s, \sigma_k) \, \sigma_k$$

is an orthogonal projection in L^2. Moreover, since each $\sigma_k \in \mathcal{H}^2$, $R(P_n) \subset \mathcal{H}^2$. Therefore, lemma 5.2 shows that the equations

$$\dot{s}(t) + P_n \, [s(t) \cdot \nabla s\,(t)] - \nu P_n \nabla^2 s\,(t) = 0 \tag{5.14}$$

$$s(0) = s_n^0 \tag{5.15}$$

have a solution in $C^1([0, \infty); R(P_n))$ for every $s_n^0 \in R(P_n)$.

Take $s^0 \in \mathcal{H}_0^{1,2}$, and let

$$s_n^0 = P_n s^0. \tag{5.16}$$

Denote the corresponding solution of (5.14–15) by s_n.

Since $\{\sigma_k\}$ is complete in $\mathscr{L}^2$,

$$P\varphi = \sum_1^\infty (\varphi, \sigma_k)\, \sigma_k,$$

so that if $\varphi \in \mathscr{L}^2$,

$$\varphi = \sum_1^\infty (\varphi, \sigma_k)\, \sigma_k. \tag{5.17}$$

Moreover, each σ_k satisfies

$$\lambda_k \sigma_k = -P\nabla^2 \sigma_k.$$

Therefore, Green's theorem shows that

$$(\nabla \sigma_k, \nabla \sigma_j) = -(\sigma_k, P\nabla^2 \sigma_j)$$

$$= \lambda_j\, (\sigma_k, \sigma_j)$$

$$= \lambda_j \delta_{kj}.$$

This shows that $\{\sigma_k/\sqrt{\lambda_k}\}$ is an orthonormal sequence in $\mathscr{H}_0^{1,2}$. We have

$$\nabla P_n s^0 = \sum_1^n (s^0, \sigma_k)\, \nabla \sigma_k.$$

Therefore,

$$\|\nabla P_n s^0\|_2^2 = \sum_1^n \lambda_k\, |(s^0, \sigma_k)|^2$$

$$\leq \sum_1^\infty \lambda_k\, |(s^0, \sigma_k)|^2$$

$$= \|\nabla s^0\|_2^2, \tag{5.18}$$

by (5.17).

Now, lemma 5.3 shows that

$$\|s_n\|_{\mathscr{X}(a,T)} \leq \frac{c\, \|\nabla P_n s^0\|_2^2}{(1 - cT\, \|\nabla P_n s^0\|_2^4)^{1/4}} \tag{5.19}$$

if T is small enough, and

$$\|s\|_{\mathscr{X}(a,\infty)} \leq c\, \|\nabla P_n s^0\|_2\, (1 + \|\nabla P_n s^0\|_2^4)^{1/2} \tag{5.20}$$

if a and $\|\nabla P_n s^0\|_2$ are small enough. However, the right sides of (5.19) and (5.20) are increasing functions of $\|\nabla P_n s^0\|_2$. Therefore, (5.18) shows that

$$|s_n\|_{\mathscr{X}(a,T)} \leq \frac{c\, \|\nabla s^0\|_2^2}{(1 - cT\, \|\nabla s^0\|_2^4)^{1/4}} \tag{5.21}$$

if T is small enough, while

$$\|s_n\|_{\mathscr{X}(a,\infty)} \leq c\, \|\nabla s^0\|_2\, (1 + \|\nabla s^0\|_2^4)^{1/2} \tag{5.22}$$

if a and $\|\nabla s^0\|^2$ are small enough. It is important to notice that the value of T for which (5.21) is valid is independent of n. Indeed, (5.19) is valid if

$$cT < \frac{1}{\|\nabla P_n s^0\|_2^4}. \tag{5.23}$$

Therefore, if T is chosen so that

$$cT < \frac{1}{\|\nabla s^0\|_2^4}, \tag{5.24}$$

(5.18) shows that this T also satisfies (5.23). Therefore, (5.19) is valid whenever T satisfies (5.24). So, then, is (5.21).

The right side of (5.21) is independent of n. Therefore, the sequence $\{s_n\}$ has a weakly convergent subsequence in $\mathscr{K}\,(a, T)$. Call this subsequence $\{s_n\}$ again, and let s be its weak limit in $\mathscr{K}\,(a, T)$. The definition of the space $\mathscr{K}\,(a, T)$ shows that $\{\dot{s}_n\}$ converges weakly to $\dot{s}$ in $L^2\,((0, T); \mathscr{L}^2)$. Also, whenever $\varphi \in L^2\,((0, T); \mathscr{L}^2)$,

$$\int_0^T (P_n \nabla^2 s_n\,(t) - P\nabla^2 s\,(t), \varphi(t))\, dt$$

$$= \int_0^T [(P\nabla^2 \{s_n(t) - s(t)\}, \varphi(t)) + (P\nabla^2 \{s_n(t) - s(t)\}, (P_n - P)\varphi(t))$$

$$+ ((P_n - P)\,\nabla^2 s\,(t), \varphi(t))]\, dt. \tag{5.25}$$

Again, the definition of $\mathscr{K}\,(a, T)$ shows that $\{P\nabla^2 s_n\}$ converges weakly to $P\nabla^2 s$ in $L\,((0, T); \mathscr{L}^2)$. Therefore, the first term on the right of (5.25) goes to zero as n goes to infinity. The second term also goes to zero, for we have

$$\left| \int_0^T (P\nabla^2 \{s_n(t) - s(t)\}, (P_n - P)\varphi(t))\, dt \right|$$

$$\leqq (\|P\nabla^2 s_n\|_{2,2} + \|P\nabla^2 s\|_{2,2})\,\|(P_n - P)\,\varphi\|_{2,2}$$

$$\leqq c\,\|(P_n - P)\,\varphi\|_{2,2},$$

since $\{P\nabla^2 s_n\}$ converges weakly in $L^2\,((0, T); \mathscr{L}^2)$. Now, the function $\|(P_n - P)\,\varphi(t)\|_2$ goes to zero almost everywhere in $(0, T)$ since $\{\sigma_k\}$ is complete in $\mathscr{L}^2$. Also,

$$\|(P_n - P)\,\varphi(t)\|_2 \leqq 2\,\|\varphi(t)\|_2.$$

Thus, the Lebesgue convergence theorem shows that

$$\|(P_n - P)\,\varphi\|_{2,2}^2 = \int_0^T \|(P_n - P)\,\varphi\,(t)\|_2^2\, dt$$

goes to zero as n goes to infinity. A similar argument shows that the last term on the right of (5.25) also goes to zero. Therefore, (5.25) shows that $\{P\nabla^2 s_n\}$ converges weakly to $P\nabla^2 s$ in $L^2\left((0, T); \mathscr{L}^2\right)$.

Next, lemmas 2.1 and 10.2.3 show that $\mathscr{H}^2 \subset \mathscr{H}_0^{1,2}$, with continuous injection. Therefore, $\{s_n\}$ converges weakly to s in $L^2\left((0, T); \mathscr{H}_0^{1,2}\right)$. Also, lemma 2.2 shows that $\{s_n(t)\}$ converges strongly to $s(t)$ in $\mathscr{L}^2$ almost everywhere. Therefore, the same argument we used in the proof of theorem 11.4.3 shows that $\{s_n \cdot \nabla s_n\}$ converges weakly to $s \cdot \nabla s$ in $L^2\left((0, T); \mathscr{L}^2\right)$. Indeed, the argument used in theorem 11.4.3 gives

$$\int_0^T (s_n(t) \cdot \nabla s_n(t), \varphi(t))\, dt \to \int_0^T (s(t) \cdot \nabla s\,(t), \varphi(t))\, dt$$

for $\varphi \in C_0^0\left([0, T); \mathscr{C}_0^\infty\right)$. Since this space is dense in $L^2\left((0, T); \mathscr{L}^2\right)$, while the proof of lemma 2.7 shows that

$$\int_0^T \|s_n(t) \cdot \nabla s_n(t)\|_2^2\, dt \leqq c\, \|s_n\|_{\mathscr{X}(a, T)}^2, \qquad (5.26)$$

it follows that $\{s_n \cdot \nabla s_n\}$ converges weakly to $s \cdot \nabla s$, since the right side of (5.26) is bounded. A consequence is that $P_n\left(s_n \cdot \nabla s_n\right)$ converges weakly to $Ps \cdot \nabla s$ in $L^2\left((0, T); \mathscr{L}^2\right)$.

Thus, we see that the sequences $\{\dot{s}_n\}$, $\{P_n\nabla^2 s_n\}$, and $\{P_n\left(s_n \cdot \nabla s_n\right)\}$ converge weakly in $L^2\left((0, T); \mathscr{L}^2\right)$ to $\dot{s}$, $P\nabla^2 s$, and $P\left(s \cdot \nabla s\right)$, respectively. On the other hand, s_n satisfies

$$\dot{s}_n(t) + P_n\left[s_n(t) \cdot \nabla s_n(t)\right] - P_n\nabla^2 s\,(t) = 0,$$

$$s_n(0) = P_n s^0.$$

Letting n go to infinity in these equations, we conclude that s is a strong solution of the Navier–Stokes equations in the interval $[0, T]$. This completes the proof of the first part of theorem 5.4. The last part of the theorem is proved in a similar way, only beginning with the inequality (5.21) instead of (5.20).

The uniqueness part of the theorem follows from 12.3.3.

Exercise

5.1 Beginning with the integrated energy equation (5.13) instead of the inequalities (5.20) and (5.21), use Galerkin's method to provide a new proof of the existence of a weak solution of the initial-value problem.

6   Smoothness and uniqueness of weak solutions

We consider, finally, the set of values of t at which a weak solution can fail to be strong (and, therefore, unique). Exercise 12.2.2 shows that in two dimensions this set is empty. It would be nice to prove this in three dimensions, but all efforts to do so have so far failed, and the preponderance of evidence seems to indicate that in three dimensions the set is *not* empty. The main result of this section was proved by Leray[1] when $V = R^3$ and by Kaniel and me[2] for bounded domains V.

THEOREM 6.1   *Let σ be any weak solution of the Navier–Stokes equations in a bounded domain V. Then, σ is a strong solution except on a set E of values of t having the following properties:*

(i)   *E is the complement of a union of half-open intervals of the form $[t_0, t_1)$;*

(ii)   *E is bounded;*

(iii)   *the Lebesgue measure of E is zero.*

Proof $\sigma \in L^2((0, \infty); \mathscr{H}_0^{1,2})$. Therefore,

$$\int_0^\infty \|\nabla \sigma(t)\|_2^2 \, dt < \infty,$$

and $\|\nabla \sigma(t)\|_2^2$ is finite almost everywhere. Let $E_0 = \{t: \|\nabla \sigma(t)\|_2 = \infty\}$, and let E be the union of E_0 and the complement of the set S_0 of theorem 11.5.5. E has measure zero since both E_0 and the complement of S_0 do (theorem 11.5.5).

Let E' denote the complement of E. If $t_0 \in E'$, $\sigma(t_0) \in \mathscr{H}_0^{1,2}$. Therefore, Theorem 5.4 shows that $\sigma(t_0)$ will serve as initial data for a strong solution on an interval $[t_0, t_0 + T(t_0))$. Let s denote this strong solution. The definition of E' and theorem 3.2 show that $\sigma(t) = s(t)$ on $[t_0, t_0 + T(t_0))$. Let $t_1 \in [t_0, t_0 + T(t_0))$. Then, $\sigma(t_1) = s(t_1) \in \mathscr{H}_0^{1,2}$, by lemma 2.5. Also, $\sigma(t) = s(t)$ in $[t_0, t_0 + T(t_0))$, while $s(t)$ satisfies the energy *equation* in that interval. It follows from this that $t_1 \in E'$, so that $[t_0, t_0 + T(t_0)) \subset E'$ whenever $t_0 \in E'$, and, therefore, that E' is the union of all the intervals $[t_0, t_0 + T(t_0))$ with $t_0 \in E'$.

To prove that E is bounded, it suffices to show that some interval $[t_0, \infty) \subset E'$. Now, $\int_0^\infty \|\nabla \sigma(t)\|_2^2 \, dt < \infty$. Therefore, given any $\varepsilon > 0$, there

---

[1] Sur le mouvement d'un liquide visqueaux emplissant l'espace. Acta mathematica 63 (1934) 193–248.

[2] In the paper cited on p. 200.

is a $t_0 > 0$ such that $\|\nabla s(t_0)\|_2 < \varepsilon$. But if $\|\nabla s(t_0)\|_2$ is small enough, theorem 5.4 says that there is a strong solution in $[t_0, \infty)$ taking on the value $\sigma(t_0)$ at t_0. Theorem 3.1 again then says that σ is the strong solution. Therefore $[t_0, \infty) \subset E'$.

It is perhaps worth mentioning that the fact that theorem 5.4 provides a strong solution for initial data in $\mathcal{H}_0^{1,2}$ and not in a smaller space is crucial for theorem 6.1.

The same argument settles the question of uniqueness of weak solutions, at least when the initial velocity is small and in $\mathcal{H}_0^{1,2}$.

THEOREM 5.2 *Let V be bounded. If $s^0 \in \mathcal{H}_0^{1,2}$ and $\|\nabla s^0\|_2$ is small enough, the weak solution with initial value s^0 is unique.*

For if $\|\nabla s^0\|_2$ is small enough, there is a strong solution in $K(a, \infty)$, and any weak solution must be identical to this strong solution, by theorem 3.1.

CHAPTER 14

A reproductive property of the Navier-Stokes equations

1 Preliminary remarks

FUNCTIONAL ANALYSIS is particularly suited to proving theorems beginning with the words, "There exists...". But sometimes matters can be so arranged that a particular property of solutions of certain equations is equivalent to an existence statement. Such a property is derived in this chapter.

The property in question is related to, and implies the existence of, periodic solutions of the Navier-Stokes equations. Consider the Navier–Stokes equations in the presence of an external force:

$$\dot{s}(t) + s(t) \cdot \nabla s(t) - \nu \nabla^2 s(t) = Q(t) - \frac{1}{\varrho} \nabla \pi(t), \qquad (1.1)$$

$$\nabla \cdot s(t) = 0, \qquad (1.2)$$

$$s(t) = 0 \quad \text{on } \partial V, \qquad (1.3)$$

$$s(0) = s^0. \qquad (1.4)$$

A reasonable conjecture is that if Q is periodic then there is an initial velocity s^0 such that the corresponding solution is also periodic. This was first proved to be true by Serrin[1] and then, using other methods, by a number of authors[2,3,4,5]. What we prove here is the following curious, and stronger, result: whatever the external force may be, given any two values of the time, t_0 and t_1, there is a solution with the same values at t_0 and t_1. This property, called the *reproductive property*, was first proved by Kaniel and me[6]. If the

---

[1] James Serrin, A note on the existence of periodic solutions of the Navier–Stokes equations. Archive for rational mechanics and analysis, 3 (1959) 120–122.

[2] Giovanni Prodi, Qualche resultato riguardo alle equazioni di Navier–Stokes nel caso bidimensionale. Rend. Sem. Mat. Univ. Padova 30 (1960) 1–15.

[3] J.-L. Lions, Sur la régularité et l'unicité des solutions turbulentes des équations de Navier–Stokes. Rend. Sem. Mat. Univ. Padova 30 (1960) 16–23.

[4] V. T. Yudovic, Periodic motions of a viscous incompressible fluid. Soviet Math. (1960) 168–172.

[5] O. A. Ladyzhenskaya, The mathematical theory of viscous, incompressible flow. New York, Gordon and Breach 1963.

[6] A reproductive property of the Navier–Stokes equations. Archive for rational mechanics and analysis 24 (1967) 363–369.

external force has period $t_1 - t_0$, the reproductive property obviously implies the solution is periodic.

In this chapter, we present a different and somewhat simpler proof of the reproductive property than that given in the reference. The proof uses the Banach fixed point theorem in the following form.

THEOREM 1.1 *Let B be a Banach space and K a convex subset of B. Let A be a continuous function on the closure of K, mapping K into itself, and such that*

$$\|A\varphi - A\psi\| \leqslant \theta \|\varphi - \psi\| \qquad (1.1)$$

for some $\theta < 1$, whenever φ and ψ are in K. Then, A has a unique fixed point in the closure of K.

We state this theorem explicitly because it is not quite the usual form of the Banach fixed-point theorem. However, it is proved in the usual way, beginning with an element $\varphi_0 \in K$ and iterating by means of the formula $\varphi_{n+1} = A\varphi_n$.

2   The reproductive property

Lemma 13.2.1 shows that every element in $\mathscr{H}^2$ is in $\mathscr{H}_0^{1,6}$ and lemma 10.2.8 shows that every element of $\mathscr{H}_0^{1,6}$ is continuous in $\bar{V}$ and, therefore, bounded.

Thus, we have

LEMMA 2.1 *Let V be bounded. Then, $\mathscr{H}^2 \subset L^\infty$, and if $\varphi \in \mathscr{H}^2$,*

$$\|\varphi\|_\infty \leqslant c \|P\nabla^2\varphi\|_2. \qquad (2.1)$$

We prove next

LEMMA 2.2 *Let V be bounded. Let $s^0 \in \mathscr{H}_0^{1,2}$ and $Q \in L^\infty((0, T); L^2)$. Let s be a strong solution of (1.1-4) in $[0, T]$. If $\|\nabla s^0\|_2$ and $\|Q\|_{2,\infty}$ are small enough, then $\|\nabla s(t)\|_2$ is bounded in $[0, T]$.*

Proof As in the proof of theorem 13.4.1, we begin by multiplying (1.1) by $-P\nabla^2 s(t)$ and integrating over V. The result is

$$\frac{1}{2}\frac{d}{dt}\|\nabla s(t)\|_2^2 + \nu\|P\nabla^2 s(t)\|_2^2 = (s(t) \cdot \nabla s(t) - Q(t), P\nabla^2 s(t)). \qquad (2.2)$$

Lemmas 13.2.1 and 2.1 say that there are positive constants c_1 and c_2 such that

$$\|\nabla s(t)\|_2 \leqslant c_1 \|P\nabla^2 s(t)\|_2,$$

and

$$\|s(t)\|_\infty \leqslant c_2 \|P\nabla^2 s(t)\|_2.$$

Let c be the maximum of c_1 and c_2. Then, for this particular constant c,

$$\|\nabla s\,(t)\|_2 \leqslant c\,\|P\nabla^2 s\,(t)\|_2,$$

$$\|s(t)\|_\infty \leqslant c\,\|P\nabla^2 s\,(t)\|_2.$$

Suppose

$$\|\nabla s^0\|_2 \leqslant \frac{\nu}{2c} \tag{2.3}$$

and

$$\|Q\|_{2,\infty} \leqslant \frac{\nu^2}{4c^2}.$$

$s(t)$ starts, when $t = 0$, in the ball in $\mathcal{H}_0^{1,2}$ defined by (2.3). We show that $s(t)$ must remain in this ball. Consider any value of t for which $s(t)$ lies on the boundary of (2.3), so that

$$\|\nabla s\,(t)\|_2 = \frac{\nu}{2c}. \tag{2.4}$$

For such a value of t, (2.2) gives

$$\frac{1}{2}\frac{d}{dt}\|\nabla s\,(t)\|_2^2$$

$$\leqslant -\nu\,\|P\nabla^2 s\,(t)\|_2^2 + \|s(t)\|_\infty\,\|\nabla s\,(t)\|_2\,\|P\nabla^2 s\,(t)\|_2 + \|Q(t)\|_2\,\|P\nabla^2 s\,(t)\|_2$$

$$\leqslant -\nu\,\|P\nabla^2 s\,(t)\|_2^2 + c\,\|\nabla s\,(t)\|_2\,\|P\nabla^2 s\,(t)\|_2^2 + \frac{\nu^2}{4c^2}\,\|P\nabla^2 s\,(t)\|_2$$

$$= -\nu\,\|P\nabla^2 s\,(t)\|_2^2 + \frac{\nu}{2}\,\|P\nabla^2 s\,(t)\|_2^2 + \frac{\nu}{2c}\,\|\nabla s\,(t)\|_2\,\|P\nabla^2 s\,(t)\|_2$$

$$\leqslant -\frac{\nu}{2}\,\|P\nabla^2 s\,(t)\|_2^2 + \frac{\nu}{2}\,\|P\nabla^2 s\,(t)\|_2^2 = 0.$$

Thus, whenever (2.4) is satisfied, $\|\nabla s(t)\|_2$ is not increasing. Therefore,

$$\|\nabla s\,(t)\|_2 \leqslant \frac{\nu}{2c},$$

and lemma 2.2 is proved.

The *a priori* inequality implied by lemma 2.2 can be used to show there is a strong solution of (1.1–4) in any interval where the hypotheses of the lemma are satisfied. With this fact in mind, we can prove

THEOREM 2.3 *Let V be bounded. If $Q \in L^\infty ((0, T); L^2)$ and $\|Q\|_{2,\infty}$ is small enough, then given any $t_1 \in (0, T]$ there is a unique function $s^0 \in \mathscr{L}^2$ and a corresponding strong solution of* (1.1–4) *that reproduces its initial value at $t = t_1$:*

$$s(t_1) = s(0) = s^0.$$

Proof Let B_R and B'_R denote the closed balls of radius R in $\mathscr{H}_0^{1,2}$ and $L^\infty ((0, T); L^2)$, respectively:

$$B_R = \{s^0 \in \mathscr{H}_0^{1,2}: \|\nabla s^0\|_2 \leqslant R\},$$

$$B'_R = \{Q \in L^\infty ((0, T); L^2): \|Q\|_{2,\infty} \leqslant R\}.$$

We saw in lemma 2.2 that there is an $R > 0$ such that the strong solution of (1.1–4) is in B_R for all $t \in [0, T]$ whenever $s^0 \in B_R$ and $Q \in B'_{R^2}$. From now on, R denotes any such number. Let Q be a *fixed* element of B'_{R^2}, and let s^0 vary over B_R. For each such s^0, there is a corresponding strong solution s. Define an operator A on B_R by the formula

$$As^0 = s(t_1).$$

If s_1^0 and s_2^0 are in B_R, let s_1 and s_2 be the corresponding strong solutions. Define

$$s(t) = s_1(t) - s_2(t).$$

Then, s satisfies

$$\dot{s}(t) + Ps_1(t) \cdot \nabla s(t) + Ps(t) \cdot \nabla s_2(t) - \nu P \nabla^2 s(t) = 0. \qquad (2.5)$$

Multiply (2.5) by $s(t)$ and integrate over V. By the usual argument involving the divergence theorem, the term $(Ps_1(t) \cdot \nabla s(t), s(t))$ vanishes. Therefore, we find

$$\frac{1}{2} \frac{d}{dt} \|s(t)\|_2^2 + \nu \|\nabla s(t)\|_2^2 = -(s(t) \cdot \nabla s_2(t), s(t))$$

$$\leqslant \|s(t)\|_4^2 \|\nabla s_2(t)\|_2$$

$$\leqslant c \|s(t)\|_2^{1/2} \|\nabla s(t)\|_2^{3/2} \|\nabla s_2(t)\|_2$$

$$\leqslant \frac{\nu}{2} \|\nabla s(t)\|_2^2 + c \|s(t)\|_2^2 \|\nabla s_2(t)\|_2^4,$$

by an argument that we have now used many times.

Since $s_2^0 \in B_R$ and $Q \in B'_{R^2}$, $s(t) \in B_R$ for all $t \in [0, T]$, by lemma 2.2. Therefore,

$$\frac{d}{dt} \|s(t)\|_2^2 \leqslant -\nu \|\nabla s(t)\|_2^2 + cR^4 \|s(t)\|_2^2 \leqslant (-c_1 + cR^4) \|s(t)\|_2^2$$

$$= -c \|s(t)\|_2^2,$$

if R is small enough. This inequality implies

$$\|s(t)\|_2^2 \leqslant \|s(0)\|_2^2 \, e^{-ct}. \tag{2.6}$$

Now, $s(0) = s_1^0 - s_2^0$, while $s(t_1) = As_1^0 - As_2^0$. Setting $t = t_1$ in (2.6), we find that

$$\|As_1^0 - As_2^0\|_2 \leqslant e^{-ct_1} \|s_1^0 - s_2^0\|_2. \tag{2.7}$$

We now refer to theorem 1.1. In that theorem, let $B = \mathscr{L}^2$, and let $K = B_R$. K is a convex subset of B, and lemma 2.2 shows that A maps K into itself. (2.7) shows that A is continuous on K, so that A has a continuous extension to the closure of K. Finally, (2.7) shows that if $t_1 > 0$, A is a contraction. Therefore, A has a fixed point. If the fixed point is s^0, and $s^0 \in K$, then

$$As^0 = s(t_1) = s^0.$$

If, on the other hand, s^0 is in the closure of K, then there is a sequence $\{s_n^0\} \subset K$ such that s_n^0 converges to s^0 in $\mathscr{L}^2$, and $\{As_n^0\} = \{s_n(t_1)\}$ also converges to s^0 in $\mathscr{L}^2$. For each n, s_n is a strong solution, and $s_n(t) \in B_R$ for all t. This means that $\|\nabla s_n(t)\|_2 \leqslant R$. Thus, $\{s_n(t)\}$ has a subsequence converging weakly to $s(t)$, say, in $\mathscr{H}_0^{1,2}$ for each t. Clearly, the function s so defined is a weak solution, while $\|\nabla s(t)\|_2 \leqslant R$ for all $t \in [0, T]$. This inequality can be used as in the proof of theorem 13.5.4 to show that s is a strong solution. Moreover, in $\mathscr{L}^2$,

$$s(t_1) = \lim s_n(t_1) = \lim As_n^0 = s^0.$$

This completes the proof of theorem 2.3.

Exercise

2.1 Using lemma 2.2, show that if $s^0 \in \mathscr{H}_0^{1,2}$ and $Q \in L^\infty ((0, T); L^2)$, while $\|\nabla s^0\|_2$ and $\|Q\|_{2,\infty}$ are small enough, then (1.1–4) have a unique strong solution.

3 Periodic solutions and stability

Let Q be periodic, with period t_1, say. Then, theorem 2.3 says that if $\|Q\|_{2,\infty}$ is small enough, there is a unique solution s reproducing its initial value at t_1. Beginning again with this initial value at t_1, we find that this (unique) strong solution reproduces its value again at $2t_1$. This argument can be repeated indefinitely, of course, and one concludes that there is a unique periodic solution with the same period as the external force. But more than this is true: the periodic solution is *stable*, that is, any other strong solution converges to the periodic solution as t increases. This result is very easy to prove using a refinement of the argument of lemma 2.2.

LEMMA 3.1   *Let V be bounded. Let $s^0 \in \mathscr{H}_0^{1,2}$ and $Q \in L^\infty\left((0,T); \mathscr{L}^2\right)$. Suppose $\|Q\|_{2,\infty} \leqslant \|\nabla s^0\|_2^2$. Then any strong solution of (1.1–4) satisfies*

$$\|\nabla s(t)\|_2 \leqslant \|\nabla s^0\|_2 \tag{3.1}$$

if $\|\nabla s^0\|_2$ is small enough.

Proof   If $\nabla s^0 = 0$, the hypothesis shows that $Q(t) \equiv 0$. Then $s(t) = 0$, and (3.1) is obvious. Therefore, we may assume $s^0 \neq 0$. Then, the argument of lemma 2.2 shows that

$$\frac{1}{2}\frac{d}{dt}\|\nabla s(t)\|_2^2 \leqslant -\nu\|P\nabla^2 s(t)\|_2^2 + c\|\nabla s(t)\|_2\|P\nabla^2 s(t)\|_2^2$$

$$+ \frac{\|Q\|_{2,\infty}}{\|\nabla s^0\|_2}\|\nabla s^0\|_2\|P\nabla^2 s(t)\|_2.$$

Now, if, for some t, $\|\nabla s(t)\|_2 = \|\nabla s^0\|_2$, we have

$$\frac{1}{2}\frac{d}{dt}\|\nabla s(t)\|_2^2 \leqslant -\nu\|P\nabla^2 s(t)\|_2^2 + c\|\nabla s^0\|_2\|P\nabla^2 s(t)\|_2^2$$

$$+ \|\nabla s^0\|_2\|\nabla s(t)\|_2\|P\nabla^2 s(t)\|_2 - (\nu - c\|\nabla s^0\|_2)\|P\nabla^2 s(t)\|_2^2,$$

by lemma 13.2.1. Therefore, if $\|\nabla s^0\|_2$ is small enough, $\|\nabla s(t)\|_2$ is decreasing whenever $\|\nabla s(t)\|_2 = \|\nabla s^0\|_2$. This proves (3.1).

Let Q be periodic, and let s be the corresponding periodic solution. In the notation of § 2, $s^0 \in \partial B_{\|\nabla s^0\|_2}$ and the proof of theorem 2.3 shows that $Q \in B_{\|\nabla s^0\|_2^2}$. This means precisely that $\|Q\|_{2,\infty} \leqslant \|\nabla s^0\|_2^2$. Therefore, the periodic solution satisfies (3.1).

Let σ be any other strong solution. Then, s and σ both satisfy (1.1–3) (but not (1.4), of course). Therefore,

$$[s(t) - \sigma(t)] + P\left[s(t) - \sigma(t)\right] \cdot \nabla s(t) + P\sigma(t) \cdot \nabla\left[s(t) - \sigma(t)\right]$$

$$- \nu P\nabla^2\left[s(t) - \sigma(t)\right] = 0.$$

Multiply by $s(t) - \sigma(t)$ and integrate. The result is

$$\frac{1}{2}\frac{d}{dt}\|s(t) - \sigma(t)\|_2^2 + \left([s(t) - \sigma(t)] \cdot \nabla s(t), s(t) - \sigma(t)\right)$$

$$+ \nu\|\nabla s(t) - \nabla\sigma(t)\|_2^2 = 0.$$

Now, using (3.1) and the lemmas of § 10.2,

$$\left|([s(t) - \sigma(t)] \cdot \nabla s(t), s(t) - \sigma(t))\right| \leqslant \|\nabla s(t)\|_2 \|s(t) - \sigma(t)\|_4^2$$

$$\leqslant c\|\nabla s^0\|_2 \|\nabla s(t) - \nabla\sigma(t)\|_2^2.$$

Therefore, if $\|\nabla s^0\|_2$ is small enough,

$$\frac{d}{dt}\|s(t) - \sigma(t)\|_2^2 \leq -2\,(\nu - c\,\|\nabla s^0\|_2)\,\|\nabla s\,(t) - \nabla\sigma\,(t)\|_2^2$$

$$\leq -2\,(\nu - c\,\|\nabla s^0\|_2)\,\|s(t) - \sigma(t)\|_2^2.$$

This inequality can be integrated in the same way we have integrated similar inequalities to find

$$\|s(t) - \sigma(t)\|_2 \leq \|s(0) - \sigma(0)\|_2\,e^{-(\nu - c\|\nabla s^0\|_2)t}.$$

Thus, we have proved

THEOREM 3.2   *Let V be bounded. If $Q \in L^\infty\,((0, \infty); L^2)$, if Q is periodic, and if $\|Q\|_{2,\infty}$ is small enough, then there is a unique periodic strong solution of (1.1–4) having the same period as Q. If s denotes this solution, and if σ denotes any other strong solution of (1.1–3) then $\|s(t) - \sigma(t)\|_2 \to 0$ exponentially fast as $t \to \infty$.*

Index